THE
A.T.
FACTOR

PIECE FOR A JIGSAW
PART III

by
Leonard Cramp

The **New Science Series**:
- THE TIME TRAVEL HANDBOOK
- THE FREE ENERGY DEVICE HANDBOOK
- THE FANTASTIC INVENTIONS OF NIKOLA TESLA
- THE ANTI-GRAVITY HANDBOOK
- ANTI-GRAVITY & THE WORLD GRID
- ANTI-GRAVITY & THE UNIFIED FIELD
- ETHER TECHNOLOGY
- THE ENERGY GRID
- THE BRIDGE TO INFINITY
- THE HARMONIC CONQUEST OF SPACE
- VIMANA AIRCRAFT OF ANCIENT INDIA & ATLANTIS
- UFOS & ANTI-GRAVITY: Piece For a Jig-Saw
- THE COSMIC MATRIX: Piece For a Jig-Saw, Part II
- THE A.T. FACTOR: Piece For a Jig-Saw, Part III
- THE TESLA PAPERS

The **Mystic Traveller Series:**
- IN SECRET TIBET by Theodore Illion (1937)
- DARKNESS OVER TIBET by Theodore Illion (1938)
- IN SECRET MONGOLIA by Henning Haslund (1934)
- MEN AND GODS IN MONGOLIA by Henning Haslund (1935)
- MYSTERY CITIES OF THE MAYA by Thomas Gann (1925)
- THE MYSTERY OF EASTER ISLAND by Katherine Routledge (1919)
- SECRET CITIES OF OLD SOUTH AMERICA by Harold Wilkins (1952)

The **Lost Cities Series**:
- LOST CITIES OF ATLANTIS, ANCIENT EUROPE
  & THE MEDITERRANEAN
- LOST CITIES OF NORTH & CENTRAL AMERICA
- LOST CITIES & ANCIENT MYSTERIES OF SOUTH AMERICA
- LOST CITIES OF ANCIENT LEMURIA & THE PACIFIC
- LOST CITIES & ANCIENT MYSTERIES OF AFRICA & ARABIA
- LOST CITIES OF CHINA, CENTRAL ASIA & INDIA

The **Atlantis Reprint Series:**
- THE HISTORY OF ATLANTIS by Lewis Spence (1926)
- ATLANTIS IN SPAIN by Elena Whishaw (1929)
- RIDDLE OF THE PACIFIC by John MacMillan Brown (1924)
- THE SHADOW OF ATLANTIS by Col. A. Braghine (1940)
- ATLANTIS MOTHER OF EMPIRES by R. Stacy-Judd (1939)
- SECRET CITIES OF OLD SOUTH AMERICA by H.T. Wilkins (1952)

THE A.T. FACTOR

PIECE FOR A JIGSAW

PART III

The A.T. Factor
Piece For a Jigsaw Part III

Copyright 2001
by Leonard G. Cramp

First Printing

August 2001

ISBN 0-932813-92-5

Published by
Adventures Unlimited Press
One Adventure Place
Kempton, Illinois 60946 USA
auphq@frontiernet.net

www.adventuresunlimitedpress.com
www.adventuresunlimited.co.nz
www.wexclub.com/aup

10 9 8 7 6 5 4 3 2 1

To our family
with fondest love

Also by Leonard Cramp

Space,Gravity and the Flying Saucer. 1954 T.Werner Laurie.

Piece for a Jigsaw. 1966 Somerton Publishing Co.,Ltd.

Piece for a Jigsaw.Reprint 1996 Adventures Unlimited.

The Cosmic Matrix.(Piece for a Jigsaw Part 2) 1999.
Adventures Unlimited.

Printed in Canada

The CMAL 1 runabout vehicle in surface effect hovering mode.

The A.T. Factor

PIECE FOR A JIGSAW THREE

THE UFO CONNECTION

An autobiographical report of previously unpublicized extraordinary UFO close encounter incidents and paranormally related aerospace technological developments

Leonard G. Cramp

CONTENTS

Acknowledgements

As the subject matter in this book consists largely of a supportive spill-over of work published in my book The Cosmic Matrix, one might think it inevitable that my acknowledgements would be, to some extent, repetitious. That is only marginally true. For instance, in addition to typing letters, articles and manuscripts etc., it fell to my wife Irene to double up as secretary, general runabout, materials purchaser and laboratory/workshop assistant for every one of the projects included in this book and more besides! These various tasks were spread over time and I would not have allowed such a state of affairs to exist had it been too heavy a burden for her. To my oft repeated protests in this respect she would reply "I don't mind and in any case if I wasn't doing this I would be doing something else"! To our dear daughters Sue and Jane who so closely mirror their mother's tolerance and support in these matters, I give my heartfelt thanks. As will be seen from the sampling in this book, there are times when such work requires more than one pair of hands, so I have been extremely fortunate and grateful for the enthusiastic help of our two incredibly gifted sons Gary and David who have always been there when needed. More often working for nothing and even hesitating to accept the meagre amounts a well intentioned, overspent sponsor could afford. In this context my sincere acknowledgement is due to Godfrey Bilton and despite the stress and disappointments, we did finally achieve success with the Hoverplane. I shall always have fond and grateful memories, like the time he arrived in his helicopter the day after our beloved dog Jason had died. Disregarding the workload, cost and loss of valuable time, Godfrey promptly took Irene and I for a slap up meal at one of the most prestigious hotels in the south of England. It didn't bring Jason back, but we felt a hell of a lot better for that thoughtful gesture. Neither has it all been near starvation and despair, as will be seen from the text, we had fun, interest and always optimistic hope.

Anthony Brown, freelance aerodynamicist for NASA special projects, became a good friend when verifying my design for the CM100 Super Car shown in Chapter 12. His dry casual Australian manner remains clear in my memory, especially the occasion when reporting his evaluation of my calculations for the Super Car centre of gravity, he said "Len, I'm afraid you're way out on your cg positioning". Then in response to my momentary look of alarm, he mischievously added, "By at least two inches or so"!

Not long after joining the then Saunders Roe Company in 1956, some of the workstaff inevitably learned of my interest and contribution to the almost totally pooh-poohed enigma of UFOs and eventually several of them proposed that the Flying Saucer Review magazine (then edited by the late Waveney Girvan) might be an interesting addition to the inter-departmental magazine club. Very soon many more of the senior staff began to take an interest, one of the first being Raymond Wheeler, then chief stressman and later to become Systems Support Director of Westland Aerospace Ltd. It was Ray who proposed my first lecture to the local branch of the Royal Aeronautical Society discussed in Chapter 4. Ray didn't take part in UFO investigations, but neither did he scoff, and it was he who first offered some scientific respectability to my personal venture into Ufology. For that and his innate kindness and assistance I have always been truly grateful. In this same context I welcome the opportunity to remember with gratitude Peter Wrigglesworth, always ready to make slides for my lectures and not least for taking part in that never to be forgotten bizarre trip to Wales in 1961 – as discussed in Chapter 4. Desmond Norman, co-designer of the Britten-Norman Islander Aircraft, who not only accepts the reality of the UFO situation, but also did me the honour of my joining his Firecracker Aircraft design team and designing a series of flying scale model aircraft for his brother Torquil's business in Scotland, also discussed in this volume....heady days!

I would also like to extend my gratitude to all those who unhesitatingly stepped forward to offer assistance - usually urgent - when needed. Nigel Bennett who helped salvage our delapidated workshop. Bruce Giddings for his jovial reassuring support and unhesitating assistance with television interviews, in particular the never to be forgotten epic journey to the BBC Studio One demonstration. His extraordinary exploits in pioneering Ultralight and Light Aircraft has always been an inspiration to me. To Norman Stewart and Bernard Partridge who befriended me when I first joined Saunders Roe. I shall always be indebted to them for making me feel welcome when we first moved to the Island and didn't hesitate to help out with some of the test rigs. Dave Meaning who helped with the initial photographs and photocopying. Peter and Jean Thorne for their unfailing kindness and encouragement. John Bailey and Chris of JCB Imaging for their interest and endless photocopying. Martin Ingram and all those who with the passage of time I may have inadvertantly overlooked, I again extend my apologies and my grateful thanks.

Foreword

When structuring the manuscript for my previous book The Cosmic Matrix, I considered it necessary to provide two reinforcing elements for the main objective. However limited available space presented something of an impasse which required a compromise solution in which only token slices of the two elements could be usefully employed, leaving the remainder to be issued as a secondary supportive book, which would be of general interest in its own right. This resulting compilation has been entitled The A.T. Factor which includes some of the author's experience in the UFO related field together with some precognitive excursions of a technological kind.

In the Cosmic Matrix I endeavoured to identify the phenomenon of precognition with a colossal cosmic grid system as postulated by Antony Avenel in his Unity of Creation Theory of the 1940's. I pointed out that any preview of future events - whether in the form of waking visions or in dreams - can logically be regarded as a form of time travel and furthermore this is probably a normal function in the matrix system. I stated that nowadays I can reflect soberly on the very clear precognitive dreams I have experienced, moreover I urged the reader to accept that the main directives in that book i.e. a method of producing energy, gravity control and propulsion, should not be excluded from the few reproduced verifiable, precognitive examples shown therein, tentatively designated by an advanced time factor (ATF measured in years).

The more extensive list shown in the following pages may serve as an illustrative narrative to reveal that it has taken a lifetime of devoted involvement along a natural progression of investigation into VTOL aircraft, rockets, space flight, to gravity, starships and energy. Indeed at one time several of the author's Royal Airforce friends whimsically suggested that this present book could legitimately have been titled with the well known RAF motto, Per Ardua ad Astra, 'Through adversity to the stars'. No doubt the A.T. Factor will be representatively adequate. It is hoped the inclusion of this *catalogue* of work will not be misconstrued as an 'I did it first' directive, the main issue at stake in the Cosmic Matrix is far too important for that kind of infantile recreation.

Among other things the reader is offered the opportunity to become aquainted with the almost unbelievable tendency for bias where something different or 'new' is encountered. Yet this is the mere tip of this particular iceberg in which intrigue, sabotage and theft lurk just below. It is hoped that this account of my experience of some of these things may help to account for the chosen format of this book.

For the technically inclined reader, it would be remiss not to add, that all the portrayed vehicles and ideas generally represent present 'state of the art' proven technology. In other words they are not merely pipe dreams or pie in the sky ideas and many have been built and tested. The apparent diversity is due to the technological compromise; that is,they fulfill specific roles and needs ranging from cost, noise levels, fuel economy, structural weight, 'off the shelf' hardware availability etc., to name but a few.

Wherever possible the reports have been presented chronologically, but this is difficult to achieve due to inevitable overlap of the work dictated by the need to eat! Also it is fair to point out that the following collection of photographs and drawings is a representation of projects initiated, designed, built (from scrap wherever possible), tested, funded and carried out at grass root level. Often with the help and encouragement of two talented sons since they were very young and always with the aid of an uncomplaining workmate and wife.

During the encompassing years there have been many other ambitious projects, including space-drive gyroscopic systems and others too many to list here. Those that are have been chosen chiefly due to their more topical merit.

As with its counterpart the Cosmic Matrix, in fairness to both the publishers and readers of this book, it should be stressed that the decision to publicize its contents is the author's and because it is truly mine I reserve the right so to do, However it may not be generally known, that particularly in the aerospace industry, the 'state of the art' technology, when published, can be anything up to ten years old. In other words, it can be taken as a certainty, that in some parts of the world there is work currently in progress on advanced technology which is so secret as to render its acquisition well into the next decade or so , depending on the degree of its advanced status. Therefor, as stated elsewhere, in this context the publication of some of the author's advanced designs is in effect a peek behind the secret status curtain and it would be correct to interpret such an otherwise generous action as being a measure of the totally encompassing nature of my findings in transport/energy technology as portrayed in the Cosmic Matrix.

To some, the inclusion of the paranormal aspect in this book may seem somewhat incongruous. They might feel that such occult matters are surely the province of occult literature. However it should be noted that it is precisely due to such isolation that the so-called paranormal has hitherto been largely neglected by scientists. All too often such information has been inclined to be centered on *non* technical matters and therefor more comfortably read by occult sympathizers. In this book however paranormal effects *do* have a technical content, the significance of which might not otherwise have been obvious to some occult readers, but more amenable to the technically inclined, or vice versa. Whether the reader be scientifically inclined or not, it is the precognitive aspect of this report which is of paramount importance. In this respect it should be noted that the included information was available long before the advent of the now popular TV series, the X Files.

This story begins then in the first chapters which have been largely devoted to the author's early introduction to the world of the paranormal, at times related to the UFO phenomenon, that has no doubt influenced my thinking for a considerable part of my life. With the exception of several time slip cases, the majority of the work has been drawn from my own experience. As I have reason to suspect that much of the UFO phenomenon has a time related basis,I have been more liberal with that kind of the phenomenon with which I have been involved. Therefor the publication of this information has inevitably incurred an autobiographical penalty for which I ask the reader's indulgence. I am additionally prompted to emphasize this by an amusing occasion when, having failed to elicit an impartial useful comment from an interested publisher friend, I had finally tried to jostle him into action by saying "Tell me John do you think there is too much about me in all this"? To which without hesitation or batting an eyelid he blandly said....Yes! In this context I should perhaps add that if at the conclusion of this volume any reader should be apt to suspect that the author carries an over sized ego, or equally oversized proverbial chip, then, as I am sure those who *really* know me will unhesitatingly testify, that is not only erroneous but regretful, for it would imply that they have entirely missed the point. Indeed it is hoped a concise summary of the included autobiographical material will reveal an adequate justification for the stand I currently take concerning mankind's questionable ability to accept and profitably employ the technology I have been shown and outlined in the Cosmic Matrix.

Finally it is interesting to note that being a technologist rather than an occult or UFO specialist, the author is insufficiently well read in that literature to have been influenced by the experiences of other writers. Therefore readers may confidently accept any similarity of views expressed in this book in a purely corroborative light.

PART ONE

Curriculum of the Fortean Kind.

I
Prelude to Precognition

From the beginning there are three important contributing factors which should be established. First, the reader may rest assured this author has always entertained personal paranormal experiences - as well as those of others - with the same degree of tenacity and exactitude with which I have devoted to engineering problems of the 'nuts and bolts' kind. Secondly, due to considerable associated experience, I could no more deny the existence of paranormal effects, than I could deny the existence of the sun. Despite this I know some occultists and mediums have suspected me as being a paranormal sceptic. This is not only quite untrue, it is misleading, for while I accept paranormal effects I have also learned to have reservations towards some who claim to produce such phenomena. To use the analogy of the telephone, one might have no doubt about the authenticity of a *transmitted* message, but the *quality* of the reception sometimes leaves much to be desired.

The third , and most important contribution, is the fact that, as all our experience is an intrinsic function of time we are naturally inclined to believe that time is a singular constant. Whereas in reality it might be multiple and even variable. In this context it is interesting to note that many out of body (OBE) percipients have claimed a total recall of their lives during an experience which in *normal* time took place in a few minutes. The following example which occurred to me illustrated forcibly that in certain circumstances the percipient/time relationship *can* in fact differ from the ordinary day to day time experience.

Having damaged a tooth in a diving accident, required an extraction under general anaesthetic. During the time I was unconcious I had a quite real sensible dream in which on a lovely sunny day I was in an unfamiliar place which going by the style of the houses could have been Spain.

During this time I met and had a drink with people, sat talking with them, went for a walk, visited shops and other places of interest etc. Certainly I spent a few hours there. It was in no way a hurried or mixed up affair, indeed I must emphasize the whole experience was pleasant and perfectly normal. So much so that it was with quite a shock to wake up and find myself still in the dentist's chair. Rather bewildered I looked at the dentist and managed to mutter "How long did that take?" He glanced at his watch and said "Oh, a little under fifteen minutes or so!"

Here it is not so much the *substance* of the dream itself in which we are interested, but the fact that in certain circumstances the latent faculty we appear to have with which to *speed up* certain stimuli. It will help to bear this in mind as we proceed.

Nocturnal beginnings

With regard to the first of the forestated three contributions, if certain agents were endeavouring to direct me in paranormal matters, then they would have been well aware of the fact that I would take some convincing. Therefore they would have to select a contact methodology accordingly. Taking as detached a position as I now can, I have to say that for me this appears to be the case, as I hope the following examples of *shared* paranormal instances will show. I would ask no more of the reader than to accept that they are absolutely true. As with the remainder of this report I know these things occurred and for that matter still do.

In the introduction to the Cosmic Matrix I was persuaded to include what to me was probably my first UFO related incident, which occurred when I was at the impressionable age of seven. This has been repeated here to provide a degree of continuity.

The Lady and the moving
star at Fyfield

In 1926 the health of my elder brother had suffered badly due to the lethal London 'smog' and he had been sent for a year to a country convalescent home known as Fyfield in Ongar, Essex. However he became so homesick the treatment was counter productive, so in the winter of 1927 it was agreed that I should be sent there as remedial company for him. I was to remain there for nearly a year which may have been the most significant time of my life.

Fyfield Boarding School. Note the clock tower, water tank room window and the dormitory windows immediately below. The author's bed was positioned opposite the middle pair.
'Reproduced by courtesy of the Essex Record Office'

At that time I was one of the youngest kids at Fyfield and among other things I found the long dark dormitory very scary at night. This no doubt exacerbated by a sinister rumbling noise created by a wind driven well pump supplying water to an overhead tank which was situated in the clock tower above my head! I well remember the beds were positioned head to foot on either side of the room facing the windows which ran the entire length of the building. My bed was situated furthest from the windows on the opposite blank wall. I can remember vividly being woken one night by the hushed, excited tones of some of the boys in a group silhouetted by what seemed to be moonlight. Puzzled by what was going on, I caught the odd remark which to this day I remember; "Cos its not a star its moving". Now wide awake, I then became petrified when John D, one of the lads whose bed head was opposite the foot of mine, appeared to rise slowly and with head thrown back and arms outstretched, stared at something above me. I thought John was sleep walking until I realized that somehow he was gliding toward me to alight on the foot of my bed! Then in a voice filled with wonder he said " Len, Len, look at that beautiful lady". I immediately responded by diving under the bedclothes and can't remember anything more of this bizarre episode except the odd remark the following day concerning the bright star and John's sleepwalking, about which he knew nothing.

Perhaps it was simply because I was one of the youngest kids at Fyfield, but I can still remember a resident nurse, who to my embarrassment seemed to take particular interest in me in the drawing and handicraft lessons.

After returning home to London in 1928 another memorable event occurred. I had been smitten with violent toothache culminating in a visit to the dentist and an extraction. Having the afternoon off from school and wanting something to do, I used the white card from a shoe box on which to sketch. To me it was only a drawing of a sleeping baby, but it apparently proved to be of more interest than that to my proud parents. I vaguely recall some indignant talk and fuss, and something to do with my school. But what does remain with me clearly is a weekend - a Saturday - I am told, when my parents took me to the school where cleaners and several people, including the arts and crafts teacher were gathered. Quite bewildered, I was given paper and pencils, ushered into an otherwise empty classroom and while the adults took station outside with cups of tea, I was told to draw 'something!'

I remember how puzzled and embarrassed I felt by the behaviour of the adults and the 'there we told you so' look on the faces of my parents. But of no less interest within the context of this book, is the familiar memory of the young lady arts and crafts teacher who provided me with free drawing equipment and more than usual attention. For ages afterwards it was difficult for me to differentiate between that lady teacher and the nurse at Fyfield. Eventually I learned what this fiasco was about. It was said that much to the mystification of those concerned, before the visit to the dentist, with a pencil I was just another ordinary kid, immediately after I was drawing like an expert. Such accelerated development is not unknown among Ufologists where others have had similar experience. In fact a similar 'overnight' development occurred to our eldest son Gary and we still have some of his original drawings from when he was the same age as myself after the Fyfield event. So remarkably professional are these, I have now and again shown them to astonished professional friends. Yet another instance of rapid development concerned a colleague who before a severe car accident was a very capable technical artist. Immediately after recovering from the accident the acceleration of his ability was bordering on the miraculous. Today he is recognized as one of the world's leading technical illustrators.

Tempting as it is I will resist the inclination to offer in depth analyses for these instances here and leave that to the intuitive sense of the reader, save to add, with the benefit of hindsight the Fyfield episode does have overtones of paranormally related activity. However, particularly within the context of the subject of this book, it is interesting to note that long before the advent of jet propulsion and rocket motors I can recall that as a youngster I was more inclined to play games involving aircraft and contraptions which floated, telescopes and journeys to the stars, rather than the 'cops and robbers' games more in vogue in those early days.

Over the years I have heard of and met many people who have had similar experiences of this kind, some of them publicly known and others not so, and in this context it is intriguing to remember that it was George Hunt Williamson's group who claimed to have received a space message via an otherwise switched off T.V. set during which the statement was made 'To the apples we salt away, we return'. I have to confess that for many years I was rather sceptical about this and other similar stories, but today I am a little better informed.

Of the many who have experienced this kind of phenomenon not least is the internationally known Uri Geller, whose seeming UFO related personal experience was told in his best selling book My Story .

Early glimpses of future aircraft

There was the day when I was about ten and had been kept behind in an empty classroom during lunch time break as a punishment for 'lack of concentration' about something or other. I actually enjoyed the rare opportunity to use such a large room for myself, promptly made a folded paper aeroplane and took delight in bouncing it off a polished desk on to an equally polished wooden floor. However there was something missing, in my minds eye I could see and feel such a small model sliding over that desktop, but this one dropped *vertically* before it bounced gently on its way. As shown in Chapter 9 it would be more than fifty years into my future before that imagery experience became manifest!

Not having access to many books and libraries, I often dabbled in exciting experiments which to me were pet discoveries, only to be disillusioned later to learn that I had unknowingly been repeating work which learned gentlemen (including Michael Faraday) had made famous many years before.

The Rainbow on the wall

By the time I was twelve the fascination for all things celestial had developed into deep inner debates about life in general. Who are we, where did we come from, was there a God etc?. I was inclined to feel that, to some extent, everybody thought like that. But I eventually learned that it was not so. So deep could this state of reverie be that on one occasion - I am told - I passed my mother on the pavement without seeing her, very much to her distress! Then on one memorable occasion, while in one of these philosophical states of mind, reclining on my bed feeling tired and close to sleep, for some strange reason - which may now seem absurd - I had the thought 'what about some physical demonstration to confirm some of these lonely daydreams. Such as, if indeed there *is* a God let a light come across the wall'. This was accompanied by an arching swing of my hand which *did not touch the wall.* Immediately a rainbow of light about four inches wide manifest and I was reassured and not a little scared!

After all the intervening years I could justifiably ask myself - as the reader may indeed be thinking now - 'did I dream that' save for the fact that my brother in bed in the same room shot up and said "Len...did you see that?."

Lest some readers be inclined to feel that such thinking was morbid or unhealthy for a young lad, let me hasten to assure them, far from it, I found it stimulating, a kind of deja vu, as though I wanted to remember something vaguely familiar. What is more I am inclined to believe that to a certain extent sometime or other most of us do.

As I grew older I tried to repeat this experiment to no avail. Since those early times I have of course anticipated that others would suspect this was merely an electrostatic phenomenon. But if so, how? For my hand had been a considerable distance from that wall. Moreover it rather beggars the question, how does one electrostatically produce a clearly *defined* rainbow spectrum? If on the other hand this effect can be labelled paranormal, then I count my blessings and place alongside some of my precognitive experiences of the type which have been verifiably chronologically published and shown in this book.

Recurring precognitive dream. 1

Due to the interrupted schooling at Fyfield, by the time I was seventeen I was academically illiterate, but I could draw well and this helped to get me a job as a trainee commercial artist. But I could also play the piano and guitar sufficiently well to win a couple of talent contests, with my sights set on the theatre as a livelihood.

In the summer of 1935 I experienced an unusual recurring dream, in which one of the most important ingredients was linked to the weather. Still living in London I and a few friends formed a small group and we were scheduled to play at a local street party in celebration of King George V's silver jubilee. Similar parties had been arranged in London and country wide.

For the occasion we designed our own costumes and my sister had agreed to make mine for me. The material was a flamboyant bright pink and the style was Mexican with large flared trousers. One night as the time drew near and excitement was rising, I dreamed that I was running along a flooded pavement in pouring rain towards the home of one of my friends, when I was suddenly flung flat on my face. The second time I had the dream I remembered the first occasion, and this occurred several times. Each time I would wake up in a sweat.

Weather forecasts in those days were not of the best and on the day of the party all the streets and decorations in London were drenched and spoilt and much to our dismay the whole public celebration was cancelled and all our efforts wasted. Returning disgruntled to my home I received the message that alternative arrangements had been made at our local school and in my resulting haste to get to my friend's house and pass on the good news, my foot caught in the flared trouser bottoms and I fell. It was not only that now my suit was completely ruined and I was suffering no small degree of humiliation, but it was the icy shock of dream world becoming reality which helped freeze me to the spot. Some might be tempted to believe that in this instance, due to the recurring dream, I may have influenced myself to have the fall. But isn't it stretching this alternative a bit far to suggest that the rainfall, all over the country, was a mere coincidence?

The future in a tea cup

Granted that precognition can sometimes be initiated in the dream state, there is compiled evidence to show that it can also be triggered by deep concentration. .

In the summer of 1938 my young wife to be and I were having tea with my sister when one of her friends named Gwen paid us a visit. We learned that she had a local reputation for 'reading tea cups', so for a bit of fun we joined in and let her read ours. Looking at my fiancee's cup Gwen said she 'saw' that we would be married and have children, two of which would be twins. (This was an accurate forecast, we did marry and we did have twin girls) Then reading my cup Gwen seemed to be rather puzzled at first, but then amazed me almost to the point of laughter, when she next said that I would become well known in the future and it had something to do with engines. Now in view of the fact at that time I was a commercial artist, a complete dunce at everything else and in particular regard to the fact that, as Gwen knew, I had only recently been made an attractive offer to become a professional entertainer in the theatre to which I was devoted , this prediction would hardly have been the result of a wild guess on her part. Indeed it was remote in the extreme to ever become true, but as will be seen later on , *every word of it did!* However Gwen went on to say that what had puzzled her was the fact that   as she looked at me she could see *engines* in the air above and going round and round my head, but she couldn't understand how engines could fly and be *circular in shape!* As stated, along with all the accounts in this book, whether psychically or technologically engendered, this account is authentic as my wife, sister and family often testified.

Talent contests and
wings of song

During those last few years I had shared my passion for aircraft with the opera, I practiced continually and was able to make a reasonable rendering in the tenor range, using my guitar for accompaniment. On summer evenings my friends and I would open the parlor windows and give impromptu arias to an assembled audience consisting of neighbours in the street. Heady days!

Meanwhile the talent contests had continued and on one unbelievable evening - whilst in the very act of introducing me and adjusting the mike - the theatre manager whispered in an aside, had I got an evening dress suit as he had a hunch I was going to need one, for a talent scout from a London circuit was sitting in the front seats to hear the show! Apparently he was favourable impressed and asked later if I would be willing to accompany a repertory troup who were doing a three weeks play at a Woolwich theatre in London. He thought I needed this exposure to get over my acute nervousness and develop some 'stage legs'. It worked and helped a lot. But of no less importance, how could I have known that one day I would find such public exposure helpful when giving future lectures ?

Several weeks later in September 1939 I received a contract (which I still have to this day) offering me an attractive salary and a permanent job touring the London theatres. I was to start on Monday September 4th, however Adolph Hitler had something to say about that, for on Sunday September 3rd WW11 was declared and all cinemas and theatres were closed for the duration!

Like Gwen, I do not consider such outpourings of talents to be anything *special* . Indeed I am inclined to believe most people possess an innate ability to do all kinds of things and this does not preclude being able to cross the boundaries of time on occasion. If indeed there is a *secret* process in all this, then I believe it lies in the ability to conjure up an intense desire to *achieve* or *know,* and by no means least, *always* accompanied in association with a fervent prayer!

The notion that in cases like mine there might have been a UFO related helping hand, is not for me to speculate and even if this were so, by no means would I be the first nor the last in this respect.

A future orientated
leap in the dark

During these years I had continued my work as a commercial artist at Lamson Paragon (at that time one of the world's largest printing companies in London), but wishing for betterment I successfully tried for a job with another small firm there. The accomodation offered by them was the usual small, cramped attic room fitted out as a studio in which no more than four people could tolerably work. My brief sojourn with them stands out clearly in my mind.

There were three older male artists in addition to myself as junior assistant, my task being to design and produce artwork for general use, such as newspaper ads.,etc. It soon became obvious to me who the senior artist was by the extremely high standard of his work, which involved some beautifully executed scraper board techniques and full colour drawings. But apart from that I found this person, a Mr. C, very different from the others. He was gently spoken and patient and always ready to offer guidance and suggestions for improvement.   While on the other hand (as nearly always to be expected in these situations) most of the others were inclined to treat me very much as the 'boy', both in the nature of the work they handed down to me and in their general manner in doing so. I noticed that at times when their conversation bordered a little on coarseness, Mr.C. did not join in, though there was no air of aloofness about him. In these circumstances his manner remained tolerant and quietly indifferent, though he certainly was no prude.

I had joined this company only a few weeks before Christmas and having worked extremely hard to successfully meet last minute rush orders, I was pleased by the acknowledgements of the staff. However I was then shocked to find in my Christmas wage packet a letter terminating my employment as being 'unsatisfactory'. I could not believe it, as from past experience I knew my work was a little more than adequate. I sat at my desk feeling completely bewildered and not a little distressed, for I had left a good secure job to go to that place. I assumed the others were aware of this by the silence in the studio and when I looked around I caught the furtive looks of several of them. When I finally exploded into vehement indignation, I was informed that much to their distaste it was the policy of the firm's senior partner to take on extra staff to accommodate the Christmas rush. Then despite all the usual promises for a secure future for the right individual etc., etc., he would *sack them*.

I came from a poor family and my small contribution meant a lot to them. I was young, extremely fit - swimming and weight lifting were my hobbies - and now I was very, very angry. Despite the others recommendations for me to exercise restraint, I was determined at least to give the boss a piece of my mind. So brushing aside the cautious hand of his secretary, I barged straight into his office and, not having met him before, was not particularly concerned at how big he might prove to be. No doubt by now he was a past master of this oft repeated situation, for as I advanced toward the dark, round shouldered, sickly looking individual hunched over the posh desk, he looked up at me, sat back and rose hesitantly from his equally posh chair. The man was a cripple. As no doubt was calculated, my temper evaporated like a pricked party balloon. I afforded this individual a few not well chosen words and left him there, no doubt still with that sickly smile on his face. I then returned amidst murmurs of "well done", "it's about time" and "sorry we couldn't let you know" etc., etc., packed my gear and left, but not before Mr.C. came over and sympathetically shook my hand and said, "I *did* try to warn you , you know". Then I remembered when I had visited the studio prior to getting the job, of them all he *had* diplomatically tried to caution me. Then with a "perhaps we'll meet again sometime" and a farewell nod from him, I went out to a somewhat bleak Christmas Eve.

War time and an
unbelievable reunion

In 1939 I enlisted in the Army and very soon found myself learning how to handle a rifle instead of a paint brush. But even then the Army found a use for such a 'facility' and had me designing and drawing kit layout procedures for the troops, and logos for our bomb disposal squad. As there was a shortage of materials, not for the first time in my life I had to make do with the cardboard from empty boot boxes!

In the spring of 1940 I was stationed for a while in the small town of Trowbridge in Wiltshire, UK, and, as newly weds, my wife and I were hoping to find a small flat where she could stay in order for us to be together before my pending embarkation.   Then another of those days I shall never forget arrived, when after having trudged unsuccessfully around Trowbridge for hours, we found ourselves in a long street of terrace houses.   Forlorn and desperate because of the approaching evening, and much to Irene's disquiet, I stood looking at the nearby buildings for a moment before making a snap decision and going to a randomly selected house, knocked on the door. Uppermost in my mind at that moment was the hope of getting some recommendation as to where we might find something suitable.   After what seemed a long daunting pause, the door opened to reveal none other than Mr.C., the artist from London!   Of the thousands of houses to choose from and over one hundred miles from where we worked together, I had chosen to knock on his door.   Despite the army uniform Mr.C. recognized me immediately, welcomed us in to meet his wife and give us tea.   That day would prove to be the beginning of my introduction to the world of aviation and eventually the domain of space, gravity and energy,

Mr.C. talked through the rest of that eventful evening. In his unreproachful manner he spoke of wrongly interpreted political issues which could have prevented WW11, he presented facts about the universe, the earth and its fitness for things and through it all my head was spinning with renewed excitement and that old familiar feeling of déjà vu.   I asked Mr.C. how he came to be in Trowbridge and he explained that as a commercial artist he had been 'requisitioned' by the authorities to take over a department at a nearby well known Aircraft Establishment as a senior technical illustrator. He suggested that I could perhaps do the same kind of work. At that time I had dismissed this as highly improbable, but a seed was planted which would eventually reach fruition beyond my wildest dreams.

We thanked Mr. & Mrs.C. and bade them farewell, we have never seen or heard of them again. A keen statistician might take the trouble to offer the odds on my finding Mr.C. like that, bearing in mind that we both could have been stationed anywhere in the UK at that time. After the first few pages of the probability rates, the results would become so high as to be meaningless to us mortals!

I have considered that perhaps in no better way can I conclude this short narrative than by adding that on Mr.C's advice we *did* find lodging for my wife and the very first evening there we discovered that our landlady also 'read the tea cups' and one of the very first things she told us newly weds was...we would have twin daughters!

An introduction to
our future

During the bombing raids on London, it was the habit of some Luftwaffe pilots, after an abortive mission, to drop their bombs on the Isle of Wight rather than waste them in the sea.   Thus it was that later in the summer of 1940 I found myself in a bomb disposal unit digging up Hitler's bombs.

The island is well known for its seams of 'blue slipper' clay and frequently the bombs were located in those areas and this usually meant we would have to dig very deep - up to thirty feet in some cases - in order to retrieve them. Having first traced the bomb entry hole and estimating its size, velocity and angle of descent, we would start digging ahead of the entrance hole, sometimes as far as twenty feet in anticipation of the bomb trajectory. In the case of a deep penetration we would have to go down in staged shelves on to which the blue slipper spoil was hurled with the aid of long bladed `grafting' shovels. (We were also assured that these might become handy as a flail in the event of hand to hand combat!).

If our estimate was correct we would first uncover the bomb's shedded fins usually at about the ten feet level, then finally the missile itself, anywhere between ten and fifteen feet further down. Ours was a motley couldn't care less crew and difficult as it now seems to believe , I know what it was like to be standing in a twenty feet deep pit with a hyped up blasé sergeant as he sweatingly stood astride an exposed five hundred pounder, triumphantly exclaiming 'gotcher you b....d, gotcher' as he clouted the obscene thing with his pickaxe!

1940. No.6 B.D.S.(Bomb Disposal Squad) jokingly referred to as the 'Bed Down Squad'. Private Cramp No.13035114 third from left,middle row.

It was while we were travelling to one of these sites to remove a thousand pound delayed action bomb dropped on a farm that 'fate' took a kindly hand. We were overtaken by a dispatch rider with updated instructions to divert to another part of the island where several unexploded sixty pound naval shells had been found. Half an hour or so later the thousand pound bomb spontaneously detonated, leaving a crater large enough in which to lose two London buses!

An important contribution to this account lies in the fact that before being detailed by the military to the Isle of Wight, I had never before been there, and although with the little opportunity afforded by the circumstances, I found the sea views, villages and countryside very appealing, the place was now rather overshadowed by the aura of war. The population was thinned by the large number of people away on active service, all the beaches had been mined and were deserted 'no go' areas. The southern cliff tops were festooned with camouflaged mock gun emplacements quickly erected from discarded telegraph poles! In fact the whole place was reminiscent of a ghost island rather than anything else. There were few passes available to us and after a gruelling week in the bomb sites , our usual 'treat' was to spend what little money we had on one decent meal at the local WVS (Womens Voluntary Services).

My stay on the island had been a brief one (some six months or so) and like many a young soldier I longed to be with my girl, family and home. Therefore the place had no special hold on me. Yet on the day of my eventual departure when at long last I was on the boat for the journey home, I still remember an extraordinary melancholy which momentarily overtook me.

I was standing towards the stern of the boat looking back at the dwindling island on the horizon, I was going home and I had good reason to be feeling glad, yet for a brief instant this was overlapped by an inexplicable sensation of sadness as if, as it were, I was *leaving my home* and everyone dear to me was back *there* on the island. Just the feelings one might have expected if my family was there at *that time*. The feeling was soon over and with slightly blurred eyes I found myself thinking 'perhaps I will go back there....one day'

At that moment I could not have possibly known that my elder brother who was stationed in the Midlands UK, would eventually meet and marry a young WRAC corporal whose home was on the Isle of Wight. The wedding would take place there and I would be invited back to the island to be my brother's best man and that primarily due to this I would be offered a job at Saunders Roe, eventually taking my entire family, wife, four children and my parents to live there. Moreover these children would grow up, get married and bless us with five lovely grandchildren!

A leap in at the deep end
for a future engineer

As rumours of our impending embarkation date drew nearer, so did speculation flourish about our possible destination and I was not overjoyed by what I heard. To go abroad on active service is one thing, to be deposited at a service store in north Iceland for the duration of the war with our motley crew was quite another! I was determined to change all that if I could.

Thus it was that when my two weeks pre-embarkation leave duly arrived, my wife and I joined my family in Norfolk, where they had been evacuated due to the bombing of London. It was there that I dreamt I was drawing aircraft components as the mysterious friend Mr.C. had suggested. Fired by this idea I devoted nearly all my leave to a complete cut away drawing of the Hawker Hurricane. How I finished it in that time I shall never know, particularly as living in a country cottage without electricity, required that I work long into the night with the aid of an oil lamp!

The work was completed and I had no time to waste with civil servants, so I parceled it up with an appropriate accompanying letter and promptly dispatched it to the Prime Minister Sir Winston Churchill himself. I briefly conveyed the situation and suggested I might be more usefully employed in the RAF. Knowing that this could bring me a whole lot of trouble, I returned to my unit very much in anticipation, with tongue in cheek and all fingers crossed!. Looking back on it now I realize my action would seem to be that of a desperate, rather rash young lad, but as we shall see later, was there something more to it than that?

Eventually I was summoned to Head Office to see the C.O. (Company Commanding Officer). Again fortune was on my side, for instead of the hard boiled WW1 veteran C.O. we had endured, he had that same week been replaced by a younger ex London Barrister, who from the word go was sympathetic to my situation. No..... we didn't receive an answer from the great man himself, but through his aide he expressed his disapproval that 'such an exceptional talent had not been more usefully exploited'. To which, predictably, the new C.O. appropriately responded for I was duly summoned to his office and immediately treated with an unaccustomed measure of army respect.

'Perhaps you should
think about that'

The following day I was mysteriously 'detailed off' to an adjacent artillery unit with instructions to visit their resident Medical Officer Captain F. Now and again over the years I have been tempted to repeat this part of the story to members of my family and close friends, some of whom have found it hilarious and others almost unbelievable, it is true none the less. The young MO looked at my Part 1 (Army Identification Booklet) which, due to short sight in my 'rifle eye', had been marked a very low health category C3, then glanced back at me then on to the covering letter asking how I felt in the process. I responded with "a fine thank you Sir", whereupon he immediately wrote out a new upgraded Part 1, closed it, wished me good luck and dismissed me! Flushed with excitement I could hardly believe my good fortune. I thanked him again, saluted and turned to the door, then unexpectedly he called me back. Hesitating for a brief moment he looked me straight in the eye and meaningfully said, "It could go *down* just as easily as it went up, you know, perhaps you should think about that!"

With my mind in a turmoil I left the office quite certain I hadn't misunderstood Capt.F. due to his reassuring nod. This presented something of a quandary for I would be unable to discuss the matter with anyone and I could hardly wait for my wife to rejoin me. The prospect for me to be transferred from the Army to the Air Force was exciting enough, but after my sojourn in the bomb disposal unit I needed a rest and the alternative of returning to civilian life as an aircraft illustrator was beyond belief.

Eventually I returned to my C.O.and discussed the matter with him. He didn't mince his words and said it was doubtful if my particular drawing ability could be justifiably used in the RAF and urged me to accept the discharge alternative. Accordingly, after a further visit to Capt.F. and adjustments to my Part 1 I received my discharge at the end of 1940. Therefore, but for the mysterious Mr.C. I would no doubt have languished my time away in Iceland for the duration of the war, and beyond much doubt I would never have found an entry into aviation. Moreover I was far from ready to spread my wings in that direction and there was much to do in its preparation.

After my discharge my wife and I rejoined our families in Norfolk and I was looking for an interim job, *any* job if necessary. One day on the recommendation of the local Labour Exchange and still wearing my old battledress replete with tatty wellies, I cycled over to a place called Snetterton Heath where an airfield was being constructed by contractors Taylor Woodrow. Perchance they would require labourers and by now I was pretty handy with grafting tools!

On which door
to knock?

Enquiries led me to a 'portacabin' type site building and I knocked on the door to be greeted by a middle aged indifferent looking individual, who eyed me up and down suspiciously, asked in a most discourteous manner "What the devil do you want?" Apparently I had misguidedly chosen the Site Manager's office, *his* higher sanctum!

Stung by this less than civil dictatorial manner, which was all too reminiscent of the army, somewhat defensively I reacted by intending to say I was a commercial artist, but for some utterly inexplicable reason this somehow became a blurted out "I'm a draughtsman sent by the Ministry of Labour". Now completely unknown to me, at that very moment the chief engineer had been trying unsuccessfully to obtain just that - a draughtsman.That site manager's manner instantly changed. Rather embarrassed he ushered me into the chief engineer's office where, scruffily dressed or not I was welcomed like a veritable gift from heaven.

Barton H. the chief engineer turned out to be a young man straight from university with a very impressive string of civil engineering degrees, though very little experience in airfield construction. He offered me a job on the spot initially to prepare scale site drawings to be used as coloured progress charts for the weekly visit of the managing director Mr. Taylor. By now, thoroughly enjoying the site manager's humble pie attitude plus the fact that I had been offered, for me, an unbelievable salary, I quickly accepted a contract for a month's trial and agreed to start work the following Monday. My wife was delighted to hear the good news and helped me prepare my all too few second hand drawing instruments with not very convincing assertions that everything would be all right. Privately I knew I hadn't a cat in hell's chance, but we desperately needed that month's salary.

Monday morning duly arrived and I was greeted by the now very courteous and friendly site manager who couldn't do too much for me. I didn't revel in this and was glad when the situation was eventually ratified. Barton H. exchanged pleasantries and after showing me around, set in motion my very first job as a draughtsman. In hindsight it was a very elementary task, but in those days without photo copiers or the like, one simply had no alternative but to literally scale everything up or down. My immediate task of scaling the site runways, perimeter track and administration sites etc.,from 1/500 to 1/2500 simply required professional scale rules and I neither possessed nor ever used one! Mercifully, excusing his absence, Barton H. then left the office to do some site leveling work and I made use of the time preparing the drawings, arranging my polished second hand instruments and drawing board backing sheet. Somehow it was as though I was waiting for something to happen, then I was taken aback when it did.

A teacher and the
curriculum

If anyone had ever told me that I would professionally meet and befriend an archetypal civil engineer straight out of the Boys Own brigade, who had built railways and bridges worldwide, I would have been most incredulous. Yet in the very moment when my little world was a very lonely place, in walked Mr. Tollast! Complete with strong sun etched features and an old broad brimmed hat and fascinating stories of jungle exploits etc., as he incessantly rolled cigarettes with tobacco stained fingers. I discovered he had designed and built all kinds of vast engineering projects and later he would tell me yarns about incredible sewage and drainage plants in which, as a youngster his passing out certificate required him to be the first to drink a glass of purified water from the effluent discharge system of his own design! With a glint in his steely eyes he would say *"that* is engineering". Later I would discover he was also a brilliant natural artist which was reflected by his draughtsmanship. I would never have believed a drawing of a sewage plant could look so beautiful. Yet despite Mr. Tollast's vast experience and capabilities, beaurocracy had given him a post in a subservient position to Barton H. Needless to say the two men did not get on very well.

At my first meeting that morning Barton H. had requested to see some of my work and I had selected a few of my pictorial designs for shop fronts which I had created as a commercial artist. Before leaving the office that morning he appeared to have been favourably impressed and satisfied with this work and the subject reverted to the job in hand which fortunately for me was going to require the forestated colour treatment. Extraordinarily enough some of these sketches were still laid on Barton's bench on the arrival of Mr.Tollast which immediately sparked a rapport between us. In a way I suppose we were both 'outsiders'. Within a couple of hours Mr.Tollast had unburdened himself a little, he had found a sympathetic ear and so had I. He laughed at my daring predicament until the tears ran down his face and responded by producing a couple of his own timeworn scale rules, and on the spot gave me my first of many elementary lessons in fundemental engineering drawing. By the time Barton H. returned I had almost completed my first job, if little else, my years as a commercial artist had taught me how to be fast!

With Mr. Tollast's kind lunchtime help, within the month's trial period I had completed all sorts of graphs and work status charts for dig, scraping, fill and concrete etc. The progress charts were coloured up each week to the satisfaction of the visiting managing director Mr. Taylor, moreover perhaps it was inevitable on more and more occasions I was asked to help with site levels, take the data back to the office and draw up my own sections. But this required some basic arithmetic so in the evening my wife (who had received excellent results at grammar school) helped polish up my multiplication tables. My enthusiasm knew no bounds. Sometimes ideas I had as a lad about flying came racing back as I strolled across the wind swept acres with a 'dumpy level' under my arm and a youngster, fresh from school as my 'chainman' walking by my side. All this was very heady stuff to me, so perhaps not surprisingly having passed the first month's trial and done a fair bit of overtime, one morning I strolled into the chief engineer's office and asked for a salary rise! Barton looked at me for a moment then said "perhaps next time round Leonard, meanwhile your present salary isn't too bad is it, and in any case you're not a first class draughtsman are you?" I left feeling a little deflated yet at the same time delighted, for within a few weeks I had risen to the dizzy heights of recognized draughtsmanship.

As the months went by I would eventually discover another reason for landing on my feet in this fertile learning ground. Apparently it was all too easy for some of the engineers - working all hours on site after site - to get bored, with what to them was after all glorified road building, resulting in the tendency to attend clandestine pheasant shooting sprees now and then. A situation which was known to Mr.Tollast, over which regretfully he had no jurisdiction.

I was to stay in this job for two and a half years and thanks to the patience of my dear wife and Mr. Tollast, by the time he had long since left and I had been transferred to another site, I had become proficient with dumpy levels and theodolites, helped set out runways, perimeter tracks, profiles, levels and not least one of the most memorable enjoyable efforts, by means of a closed traverse - which took us over miles of fields, streams and woodland - setting out the basic centre lines of the now famous three runways at Snetterton Heath airfield in Norfolk. There were sewage and storm water drainage systems , utility and administration blocks, experience that one day in the future would enable me to augment my income as a freelance architect.

Extraordinarily enough this kind of engineering experience proved to be an imperative adjunct to my overall appreciation of the laws of nature including gravity, which was often neglected by most aircraft engineers of that era. By contrast it was good to be outdoors under a blue sky, breathing the fresh scent of the Norfolk air. One could think, feel and dream easier there than is normally possible in a noisy, artificially lit drawing office. Thus it was that but for one helpful aging engineer and so called 'fate' none of this broader 'education' would have been possible!

As we lived in the country with little access to libraries and books on elementary physics and mechanics, I frequently reasoned an idea through with the aid of legions of 'thumb nail' sketches. The number of occasions when I thought I was inventing a new idea or some gadget, was extraordinary. Some of this work extended into the nature of light, electrostatics, electromagnetics and centrifugal force etc. Then later when we moved into more urbanized parts of the country with access to libraries, I discovered with mixed feelings that it had all been done before.

Although all this was damned hard going and often terribly frustrating, today I can look back and see that once over the beginners hurdle, having little knowledge as taught at elementary school, can be a distinct advantage in many ways. But of no less importance to the main directive of this book, I now understand that such a fresh uncluttered mind can be - and often is - used as fertile land for the nurture of ideas of things yet to come.

By this time it seems my sojourn in the 'deep end' was over, I was ready for the aerospace venture to begin, but how? For by now I was happily reconciled to an extremely attractive career in civil engineering. But it seems there was yet another pattern in this intricate weave.

Promotion and dismissal

One night I was working alone in the office at an otherwise deserted and nearly completed airfield. My father and younger brother had come up with an idea for a quicker method of mass concrete distribution and I was drawing it up for them. I heard footsteps coming along the corridor and in came Mr. Taylor and several engineers. Having seen the work he showed great interest in the idea, but no less in myself it seems, for several days later I was offered a transfer to the London head office as chief draughtsman!

Now this was a young man's dream come true situation and I was delighted, although it meant I would be living further away from home to endure the bombing raids on London. However I took up the appointment just in time to help in the finishing stages of the now historically famous concrete 'Mulberry Harbour' barges used in the D.Day invasion of Europe.

But my elation was soon dashed, for instead of the interesting work I had been engaged on in East Anglia, I now found myself with literally nothing to do except keeping an eye on several equally bored youngsters who made my life hell to boot! Finally I met Mr. Taylor and asked if I might return to outdoor site work and gave him my reasons for making this request. I must have caught him in a bad mood, for when he saw I was adamant he exploded "People like you should be working under Hitler." As it was highly unlikely he had ever held a rifle in his life, let alone dig up unexploded bombs, and as I thought I had his interests at heart as much as my own, I was not amused, voiced my contempt and resigned on the spot.

A multi faceted precognitive dream

In 1942 while working at Snetterton Heath and our first born son Gary was only a few months old, I had a very vivid disturbing dream. In it I was lying in bed having been wakened by someone coming up the stairs to our bedroom. Typically of dreams, while still lying in bed , I was also looking down the stairs at the ascending figure. He was broad shouldered, of middle height, fine features and a shock of golden red hair. He seemed to be ignoring me and I was rather shaken.

The next instant I was standing precariously on a narrow high up girder which was part of the open metal framework of a new building. The building was on fire and inside this structure I knew baby Gary was lying. Somehow I had to reach him.

Now although I have always been fascinated by space and flying, I can get something akin to vertigo just thinking about heights. Irrationally, sitting in a high flying helicopter isn't even comparable to looking over the side of a high bridge! Therefore in the dream, despite my desperate need to reach Gary, I was rooted to the spot, petrified.

Enter the same stranger. He glanced in my direction for an instant, perhaps indifferently, before stepping out along the narrow girder to retrieve my helpless child. With great relief and gratitude I took the baby from the stranger, inadequately blurting out my thanks. Whereupon the scene became calmer, the flickering firelight on his features turned out to be merely the glow of a beautiful sunset. The stranger gazed at it for a moment before he said "The sky is the most beautiful golden yellow", and he was gone.

The dream made a great impression on me, but it didn't cure my fear of heights, because sometime after that an engineer friend Frank invited me to join him on an ascent to the top of a very lofty aerodrome water storage tank. After the first landing I chickened out, let alone clamber up the vertical side of the tank where Frank stood looking down at me!

Reflecting on this event I was reminded of my dream of high structures and the possible Freudian reference to a 'golden yellow'. Yet that had not been said in a reproachful vein. Also I wondered if the two events might be related. But if so it was only part of the story, for the other half of which - as will be seen later in Chapter 11 - I would have to wait some forty years to decipher.

Working on some of the partially completed but operational airfields had inevitably stimulated ten fold my fascination for aircraft. I met pilots, air and ground crews and on occasion flew with them as emergency 'active ballast' on test flights. I could now think and speak the language and study more meaningfully. This integration was quite natural. If you were working way out on the site somewhere and the pilot of a damaged Mosquito aircraft had selected your particular airfield for an emergency landing and in so doing accidentally ran over one of your airfield trenches thereby ripping off the aircraft's tail wheel (resulting in a near fatal crash) you simply left your site work and got over there...quick. In fact this was such an event in which having bounced nearly one hundred feet in the air, the tail wheel 'Pogoed' across the airfield like some banshee from hell and landed up a few feet from my office window! Despite this close interrelation, when I eventually left Taylor Woodrow, I fully expected to be redirected to another similar program in the site construction industry in the capacity of junior civil engineer which I then held. I had no idea that this would not be so.

In the winter of 1943 I was once more back home in Norfolk without a job. I went to the local branch of the Ministry of Labour to apply for temporary casual work until another post could be found. But apparently the Ministry had other ideas, for no sooner had they read my particulars and the mention of my aviation related interests, I was promptly dispatched to the De Havilland Aircraft Company at Hatfield, UK, as a technical illustrator! I was extremely apprehensive about this and doubted very much if I was adequately suited for the job. Therefore having a week's grace before starting with the firm and remembering Mr. C's remarks I decided to familiarize myself with some up to date 'know how'.

Now I could have selected any one of the thousands of branches of the technology, but thumbing through one of my books, a page seemingly wanted to open itself (this has happened to me many times). It carried an article on aircraft undercarriages, so in desperation and rather than do nothing at all, I swotted up on that.

On the first Monday morning after I had been shown round and introduced to the staff at De Havillands, the section leader gave me my very first job on full size aircraft. It was the *undercarriage* for the then top secret De Havilland Hornet prototype! From then on with the help of my understanding helpmate and wife, I have had to seek, study and learn the rest of what I know today. It may sound tough, but in fact it has been exciting, often occasionally accompanied by a sense of rapport as though I was remembering something long since forgotten. But also it was often very disconcerting to discover when enquiring of engineers why were things done this way rather than that, they frequently were honest enough to admit they didn't know.

As stated, in earlier days, in order to evaluate a technical problem I would portray it in picture or model form, much as an artist would, often apologizing for this basic descriptive technique and it took me over thirty years to get out of the habit of feeling inadequate or technically inferior in this regard. Until one reassuring moment when it dawned on me that not only was Einstein a member of the 'slow beginners' brigade, but he also often used such visual aids. All those lonely years I had been in good company indeed. Today I give lectures, design aircraft, aero engines, rockets, space drives, boats and houses about which friendly engineers have sometimes sought my advice so that now and again I have found it difficult to resist a self indulgent smile at the idea of how many of them will be amused by all this now. But more to the point of this narrative is the fact that it was primarily due to my *lack* of early learning, and therefore implanted bias, that I have sometimes seen things in a different light. Time will bear credence for the veracity of the ideas expressed in my books, as it has already done during the past two decades and I feel humbled for the somewhat lonely journey I consequently occasionally endured.

A superimposed
precognitive dream

One night in the summer of 1950 I had another memorable dream which has been recorded elsewhere. I dreamt I was kneeling beside a stretch of clear transparent water. My shirt sleeves were rolled up and my clenched fist was immersed in the cold water up to the wrist. Protruding from either side of my fist was a large beautiful goldfish, still alive and in the dream I was squashing the poor creature to death. The dream fortunately terminated by my waking up in a cold sweat, thinking, 'For God's sake why would I want to do such a thing?' It made a lasting impression on me and I put it down to a typical 'nightmare'. Then one *Saturday afternoon* - shortly after the dream - my wife Irene and I visited a property in the country which was approached over a small bridge spanning an attractive pond stocked with the largest goldfish I had ever seen. Peering approvingly down at them through the clear water, I was forcibly reminded of the dream, but promptly dismissed it as of little consequence. On the *following Saturday afternoon* we took our small son Gary in a rowboat on a park lake in which there was a small island. As it was his first boat trip he had equipped himself with the usual jam jar and net, just in case he could 'catch some tiddlers'

Having reached the island Gary became quite excited by the fact that several golden carp (silver in colour) were descaling themselves by threshing about near some small rocks and in the very act of lowering his small net, a carp leapt out of the water and nose dived into it, literally it seemed, gave itself up! So comparatively large was the fish it promptly burst the net and ended up entangled with head out of the bottom and tail at the top. Our first reaction was to put it back into the water, but our little son was very disappointed, this after all was his very first 'tiddler'. So in desperation Irene held the fish's head in the water filled jar while I rowed frantically to the quayside. Not really knowing what else we could do, she and Gary raced off to the boat shed and came back with a large empty paint tin, filled it with water and released the fish into it. But instantly the fish began hurling itself out of the tin on to the ground as fast as we replaced it, not helped very much by a frantic child. By this time I noticed that the fish's gills were becoming caked with loose sand and having despatched Irene off to the boat shed to find a larger container, I did the first thing that came into my head, I grasped the poor thing and in *order to clean its gills* thrust it into the lake. It wasn't due only to the chill of the water that sent shivers down my spine!

Now here we have a dream, spread over it seems two separate weekends, both on a

Saturday afternoon, both involving fish, one a *real* goldfish, the other a golden carp. One distraught son, two devastated parents, one of whom rather than *hurt* the poor creature as in the dream, was very anxious not to let it go on the one hand, but desperate *not* to crush the life out of it on the other! Was there a happy ending? Well Irene *did* find a large saucepan,which, with another net stretched over the top, contained the fish until we arrived home where it was transferred to a large tank for a while so Gary could show off his first *tiddler* before releasing it into a local stream.

Such phenomenon is far from unique. Other researchers have listed irrefutable cases where future events have been scrambled over several dreams. The decoding is the problem. It helps a little if the researcher happens to be a participant. The author did not take part in the following incident neither can it be correctly described as a shared event. However as an example of waking spontaneous precognition it has a very appealing and moving content, especially as the chief participant was a small child.

'Poor pussy hurt'

In the summer of 1982, my daughter-in-law was out walking with a friend Pat and her little daughter, barely three years, in a pushchair. They were walking along a pavement which was separated from the main road by a grass verge, when suddenly the little girl became agitated and leaning out of her pushchair pointed to a spot on the grass verge exclaiming "Mummy, mummy, look poor pussy hurt". The child was so disturbed that Pat stopped and looked at the spot to see what was troubling her little daughter. Despite the fact there was nothing to be seen, the child still pointed insistently to the grass repeating "Poor pussy" The two women dismissed the matter as being a kiddy's fancy and walked on. Having completed a purchase at a local shop, they were returning when they saw a man placing something on the grass at the very spot the little girl had pointed to fifteen minutes before. The little child just looked but said nothing. Laying at their feet on the grass was a cat *which only moments before had been killed.*

It is well known that such psychic ability is not uncommon among children, but it is often apt to decline as they get older as in the foregoing instance. In which case one might reason   that such a faculty merely lies dormant, but might be triggered when certain energetic circumstances prevail. Thus the phenomenon might become of the shared variety, such as that related in the following chapter.   But as a footnote here it is appropriate to make an observation on the argument proffered by Arthur C. Clarke and others to tentatively explain precognition and synchronicity as being a statistical odds phenomenon. For the unacquainted a well known scenario is considered viz. Twin girls were born in the UK to a single mother whose circumstances deemed it necessary to part with one child for adoption. In her middle thirties the UK twin successfully traced her sister to America where she had been raised since a baby.

Subsequent checking revealed; both sisters had married men with the same christian names and similar appearance. Both husbands had identical interests and employment. Both families had children of the same gender and age etc., etc.

The statistical odds theory holds that this could happen by chance given sufficient time.   It is more typified by the popular scenario in which a horde of monkeys given typewriters and paper could theoretically type a Shakespeare play by random punching alone. In his remarkable book The Intelligent Universe, Sir Fred Hoyle points out the Universe as observed by astronomers wouldn't be large enough to hold the horde of monkeys required to write but one scene from one Shakespeare play, nor the waste paper baskets to contain the rubbish generated. To which I would like to add that to accept the idea of such numerical meanderings is one thing, but what about cases like the present author where, as shown in this book, precognition and synchronicity phenomena have occurred not just once but *many times* during one lifetime. This does rather reduce the statistical odds theory to an ever diminishing absurdity.   That such a significant compounded statistic should have been hitherto overlooked is probably indicative of the rather selective areas of study by some scientists.

2
Precognitive Links

As is well known to students of the paranormal, there are many case studies which suggest that in certain circumstances time appears to be variable. With the exception of several others, in this book I have drawn mainly on incidents with which I have been directly related. The following is one of them.

Shared waking time slips

This incident is additionally interesting in that not only is it known to the author, that it was a shared experience and would suggest both forward and backward time slips, but also because it occurred during a well known UFO 'flap' on the Isle of Wight in 1969.

On January 4th Dr. and Mrs White were driving to the pleasant village of Niton to dine with friends. They had decided to take the slower road across the Downs of the island and that night despite a full moon it was cloudy and dark. They were approaching the first hill of the Downs with moonlit chalk pit on the left and fields to the right of the road. With the nearest farmhouse some miles away, this is usually a lonely spot so they were not prepared for what they saw. Mrs White said, "The fields appeared to be covered with bobbing lights, as though many people were moving about there". Although initially surprised they had assumed that shepherds were working their sheep. Then having arrived at the top of the hill the couple were about to drive down when they realized that the fields to their right were also ablaze with coloured lights "Like a great modern city". Astonished they stopped the car and stared in wonder at the unaccustomed myriad of twinkling points. At that moment it had not occurred to them that the explanation might have been other than a rational one.

Ahead what they knew to be an ordinary cart track to an outlying farm, now seemed to be a *well lit city street with buildings on either side*. The lights were green, red and orange. The acute feeling of unreality was increased when they arrived at the farm track only to discover that it was now 'its usual dark and deserted self' without trace of buildings or artificial lights of any kind. By now really unnerved the Whites resumed their journey with little other than the moonlight to reassure them. They were looking forward to reaching the Hare and Hounds, a normally friendly island inn at the cross roads to the town of Newport and Merstone. They turned the corner and the inn came into view, but it was now bathed in light and 'surrounded by what appeared to be figures carrying torches' who were running backwards and forwards across the road ahead. Once again they could see the fields brightly lit and suddenly one figure stood out from the rest, a very tall man with clear cut features who ran directly in front of the car. The Whites noticed he wore a leather jerkin with a broad belt. At this point the couple decided to stop the car to enquire what was going on, it did not occur to either of them that what they were seeing was illusory until they were about twenty yards from the inn when both lights and figures suddenly vanished 'as though a switch had been thrown' and but for the usual lights from its windows the Hare and Hounds was now in darkness. The completely unnerved couple then drove on without stopping until reaching their destination. When they returned by the same route in the early morning, the landscape was normal and the strange feeling of oppressiveness had gone.

Mrs White wondered if the extraordinary experience related to past *and* future intervals of life in that area. She figured the Roman legions had once camped and marched across what was known as Vectis. All Roman camps were known to be built with at least two intersecting streets at right angles . Researchers pointed out that torches were used for lighting throughout the Empire and one might expect to find a camp adequately lit in times of peace.

Time and a little dog

The second case of time interaction occurred to Laura Lee Daniels who told the columnist Joyce Hagelthorn of Dearborn Press, Michigan, 10th May 1973, that walking home through deserted streets after working late one night, she perchanced to look up at the moon. When she looked back the urban surroundings had vanished 'even the pavement on the sidewalk was gone and I was walking on a brick path. There were no houses on either side of me, but several hundred feet before me was a thatch roofed cottage....there was a heavy scent of roses and honeysuckle in the air. As I walked up the brick path and drew closer to the cottage, I could see that there were two people sitting in the garden...a man and a woman....in very old fashioned clothes. They were obviously in love....they were embracing....and I could see the expression on the girl's face'.

As Laura wondered how to signal her intrusion, a small dog ran towards her barking.'*He was quivering all over.* The man looked up and called to the dog to stop barking....I somehow realized that he couldn't see me...and yet I could smell the flowers and feel the gate beneath my hand. While I was trying to make up my mind what to do, I turned to look back the way I had just come....and there was my street! But I could still feel the gate...I turned once again to the cottage....it was gone and I was standing right in the middle of my own block, just a few doors from home. The cottage, the lovers and the wee dog were gone.'

In this case it is interesting to note that the little dog of a past time was relating to a visitor of the future while she (Laura) was relating to its past.

A spin into the future

The reciprocal nature of the foregoing case is by no means uncommon, but interesting as such cases are, to the layman they may appear to be lacking in corroborative content. This can hardly be said about the following case in which the percipient was later able to ascertain technical verification. It concerns an experience which occurred to Air Vice-Marshall Sir Victor Goddard in the 1930s in which there can be little doubt that he interacted with the future. In an article entitled 'Breaking the Time Barrier', he describes his flight in an aircraft with open cockpit, no radio nor blind-flying instruments. While in thick cloud he suddenly stalled and went into a spin he was unable to correct. Fast losing height and aware that there were mountains in the vicinity he struggled with the controls and finally emerged at two hundred feet into 'a murky sort of daylight'.

Suddenly spotting the Firth of Forth he was able to regain his orientation. He later said

'Thanks to the railings on the esplanade, and a girl who was running in pouring rain with a pram - she had to duck her head to miss my wingtips!'

Now flying under the low cloud he made for Drem airfield (which he had visited only the day before) and after identifying the Edinburgh road, looming ahead he saw the black silhouettes of hangars. In deluging rain and dark turbulent flying conditions, he was crossing the airfield boundary when.....

'The next moment the airfield and all my immediate surroundings were miraculously bathed in full sunlight as it seemed to me, the rain had ceased, the hangars were nearby, their north-west doors opened. Lined out in spick and span order on a newly laid tarmac were four aeroplanes; three bi-planes of a standard flying training type of aircraft called Avro 504N, one monoplane of an unknown type. We had at that time no monoplanes in the RAF, but the one I saw then was of the type which thereafter I carried in my memory and identified with the Magister which became, later on, a 'trainer'. Another peculiarity about the aeroplanes on the tarmac was that they were painted bright chrome yellow. All aircraft in the RAF in 1935 were exclusively aluminium-doped, there were no yellow aeroplanes. Later, because of an alarming increase in fatal accidents at flying training schools during the first phase of the expansion of the Air Force, the needs became apparent for making training aircraft easily seen: in 1938 and 1939, more probably the latter, yellow aeroplanes became universal at all RAF flying training schools.

In the mouth of the hangar closest to me, another monoplane was being wheeled out. The mechanics pushing it were wearing blue overalls. As I passed over them, having climbed from only a few feet above ground to just high enough to clear the roof of the hangar, I must have been making a great deal of noise and, normally this would have caused a considerable sensation. Zooming the hangars, as I was doing, was a court-martial offence! It was quite certain that those mechanics must have looked up at me (had I been 'there' to them) as I flew over so close. But none of them looked up. This struck me as strange. It also struck me as strange that the airmen were wearing blue overalls. RAF mechanics had never worn anything else but brown overalls when working in hangars on aircraft. The hangar roofs, above which I was flying a moment later, were gleaming with the wetness of recent rain, but the bituminous fabric was entirely new and in very good order'

Drem airfield was rebuilt in 1939 and eventually became a flight training school. It was equipped with *yellow* painted 504N biplanes and Magister monoplanes, similar to the ones Sir Victor Goddard had seen.....and airmen were then given *blue* overalls.

From precognitive dreams
to PK phenomenon

In 1946 when our son Gary was barely five years old we had moved to a house in London. It was there that we were introduced at first hand to Psychokinetic (PK) manifestations. Fortunately for us it was to be of the benign kind. To begin with it took the form of noises; loud raps, rustling, light switches going on and off etc. We occupied an upstairs room and the raps developed into very defined rat-a-tat-tat sequence, sometimes on the floor right under my feet with nobody else in the house but us. Then the personal physical effects began. At three o'clock one morning I was wakened by a rapid fluttering between my cheek and the pillow. Not only could I feel it, but it was accompanied by a slight noise on the material (which later I was able to mimic by flicking a piece of stretched material with my thumb and finger   This nocturnal phenomenon was repeated night after night, always at three o'clock and as I awoke I became aware that it always began in the form of three close taps followed by a longer pause then three again for several times until I became fully awake. Moreover I soon realized there was an instantaneous response to my thoughts. So using the time honoured once for 'yes', twice for 'no' technique, I was able to obtain some degree of communication. Also, it became evident that there appeared to be an absorption of energy of some kind, insofar that at the beginning of these 'sessions' – which usually lasted about five minutes - the response to questions was instantaneous, the flicks and assorted noise being very energetic. Then after a few moments the potency of the flick would diminish together with an increase in the response time. I also noticed that sometimes during these sessions my wife stirred in her sleep as if dreaming. Naturally we talked about this strange phenomenon and both agreed it was of the type categorized as physical phenomena. Extraordinarily enough, it caused us no inconvenience, in fact it was almost as if we were taking part in an experiment of some kind. In fact when one of these short sessions was over and all 'power' seemed to be gone, I simply drifted off to sleep again.

Having talked the matter over with a friend, I decided on several experiments , two of which follow. I made up a sandwich comprising a small plate of thin plywood, then a layer of carbon paper with its printing surface covered by a sheet of typing paper. I reasoned, that if the effect was powerful enough to create a little noise on material, then perhaps I might obtain a printed impression of the flicks, together with an idea of the size of whatever it was causing them. Examination of this 'pressure sandwich' the following morning revealed several irregular impressions had indeed been formed.

Then we discovered that other people could share this experience, for having previously agreed, one night I woke my wife and lifted my head a little so she was able to feel the taps on the pillow herself.

A hard nut to crack

In the next development I began to receive taps on my body during the day. Almost unnoticeable at first, but gradually increasing over several weeks, so that I finally had to accept that this was due to something other than a twitching muscle. Moreover there was the 'call sign' of three initiating taps. Eventually any remaining doubt about this latter development was finally resolved when I was awakened by not only the usual three taps on my cheek, but in perfect unison on both my shins as well! Someone, somewhere had considered I was a hard nut to crack!

Now and again the phenomenon developed into little bursts of rapid pulses with some kind of coherence which at that time was difficult to identify. I also began to recognize the once for 'yes' and twice for 'no' sequence coincident with my thought patterns at the time. These physical taps were so real, on occasions I could be taken off guard by a friendly tap on the shoulder and either look round to see who it was - or perhaps feeling a little indifferent towards my invisible contact - completely ignore it only to find that one of my 'earthly' friends was responsible!

A friendly helping hand

At this time I had been employed at De Havillands for nearly a year where the nature of the work called for very careful preparation in the early stages. For an error to creep in at that juncture could involve considerable rectification, time and inconvenience for all concerned. One instance comes to mind in which I was preparing an intricate drawing of the engine installation for the twin engined DH Dove aircraft to which was fitted a three bladed variable pitch propellers. These are notoriously difficult to portray graphically and during the morning I had been interrupted by an insistent double 'NO' tap on my shoulder which, feeling rather uncertain about, I had ignored. However the taps were persistent so I very carefully went over the work again, but still couldn't find anything wrong. Feeling a little irritated by what seemed to be a false interruption, I continued with the rest of the work. The taps came again, two for NO in several sequences.

Whether or not something else distracted me I cannot now be sure, but I eventually recommenced the job and carried on uninterrupted. Late that afternoon I realized things were not working out with the work. Several key points on the layout didn't tot up. There was nothing for it but to review the whole procedure. Hours later I discovered that due to several interruptions in the office, I had indeed made an error, my unseen helper had been right all along!

By 1949 I had been working at D. Napiers in London for over a year. As discussed later, much of the work was related to secret forward projects designs and among the engineers were several who were interested to form a psychic phenomena study group, with the hope of examining levitation. I had already seen this, but had not been involved personally. After the first few meetings we obtained the phenomenon of table tilting.

Levitation

Always being my own biggest sceptic and having a healthy regard for people's frailty, I determined that this research would be conducted scientifically, but in order to make a start we should take obvious minimum precautions with regard to positioning, lighting etc. We were rewarded when the table eventually rose and proceeded to dance across the floor bringing us down on our knees laughing like 'kids at a party', seemingly aided and abetted by the infectious raucous laugh of a somewhat diminutive ex-navigator friend. Despite the fact that our party consisted of several engineers, their wives and mine, who simply wanted to examine such phenomenon, I was intrigued but still uncertain. So on that occasion, merely as an interim somewhat less than scientific but reliable precaution, I asked the group to form an unbroken circle round me and the table by clasping each others hands and touching foot to foot. The usual red light was dimmed a little and placing my hands lightly on the table top I gave thanks and as always, reverently asked for assistance.

The room now became quiet and still and because of the restriction I had reluctantly imposed I did not expect anything to happen, yet it did. Slowly, gently at first, the table edge nearest to me began to rise, my outstretched hands slipping over its polished surface. Then just as gently it lowered again. I stood there trying to relax my rigid body and slow my fast beating heart. Once again the table edge rose a few inches, but this time it was accompanied by a tilt from side to side - an unmistakable rocking as if a prelude to breaking free - causing my now sweating hands to slide towards the centre. During the next few seconds I suffered a fusion of hope, unexpected 'eureka' and disappointment, crowned by the never to be forgotten moment when, with indescribable subtlety the table rocked even more as if completely suspended. I stood there transfixed with my palms seemingly merged with the floating table before me, accompanied by a degree of moist cheek reproach for my self indulgence, when, as if in acknowledgement, the table rocked and lowered gently back to the floor.

Now I know I had not - could not - have lifted that table myself, as did those around me and I have since often given thanks and counted my blessings in that regard.

Corroboration

Shortly after this event my wife and I and two other members of the group Mr. & Mrs.

L.D. attended a public demonstration of clairvoyance at the Wigmore Hall, London, given by the world famous medium Eileen Roberts. Our seats were in the centre towards the first few rows in the hall, with our two wives sitting together between Mr.D. and myself. There was over a thousand people there that night and we had not discussed the affairs of our group with anyone. Yet during the early part of the meeting Mrs Roberts pointed towards our vicinity asking if there was a Mr.D. present, if so would he put up his hand, which he did. Her next remarks have remained in my mind, for she then asked "Where's Len?" I put up my hand, then she next said "My word haven't the four of you been having fun with the table lately!" We could only acknowledge with nods and smiles. Then still addressing herself to me she went on "You have been receiving taps on your shoulder lately haven't you" I nodded in agreement followed by "Well you have a navigator friend in spirit, who was shot down during the war and he wants you to learn the morse code so he can communicate."

The numbers game

Perhaps fortunately, most natural psychic mediums are inclined to be indifferent to the 'doubting Thomas's' of this world and in any case they can always obtain a degree of sympathetic support from their peers. But, particularly in those days, if you happened to be an aerospace researcher the situation was....a little different. Even more so if one of your biggest sceptics was yourself! I determined on a series of practical experiments. The first being as follows.

At that time my wife and I were still living at the house in the east end of London in which the psychic phenomenon had started, while the other members of our small group lived at Ealing on the other side of London an hours journey away. Both they and I were provided with identical sheets of graph paper which were numbered up to one hundred. At an agreed time the Ealing group would randomly select and mark a number. Using the one tap for YES technique, with prior agreement, my spirit helper would signify the chosen number.

This experiment was duly carried out and at a convenient time the following morning - admittedly not unlike kids at a party trick - we compared the numbers to find them identical. Now of course there could be two more obvious responses to that by lobbyists from either side of the paranormal divide viz. The sceptics would brandish mathematical probability chance factors while some orthodox metaphycisists would suggest this was merely a form of mental telepathy. Personally, with respect, I neither agree nor care, I only *know* that this phenomenon happened and has since happened to me countless times during the rest of my life.

Least some readers now be tempted to suggest that given the validity of this phenomenon, the author could be one of the richest people in the world, or something like that, I must hasten to say it does not work like that. I live my life totally independently just like anyone else. I do not and never have, lived with 'one foot in the grave'. ' This phenomenon' which has now developed more sophisticatedly, is benign and even capable of gently reproaching me in moments of aggravated temper. Or like as not ignoring me when I have been tempted to ask for help. Or at moments when nervous and feeling totally alone I have received unexpected reassurance. As with most of us there has been moments of crisis in the family when I have been assured that this too would pass. Again it has always been right, there have been gratifying extraordinary healing for animals as well as the human kind, together with revelations and predictions in the sphere of physics and mechanics in association with this comforting link, that perhaps all too often I do not deserve. I only know beyond any doubt that it exists and assists me and I am truly grateful for it. To try to take advantage of it would be wrong, whatever the source is, *it just does not function that way.*

Shared hallucinations?

Among my friends at Napiers were two other engineers who fortunately were both exceedingly sceptical about paranormal matters. Fortunately, because the following shared PK events are more evidential than they would have been had the participants held similar views. Bill H. a brilliant technical illustrator, keen motor racing enthusiast had his sights set on Brands Hatch. He and I were standing in the middle of our drawing office discussing a work's matter. The large building in which we stood was lit on two sides by very deep windows, which, being winter were closed. In the middle of the discussion someone seemed to have opened one of them causing a draught of cold air to pass by us. Instinctively I had paused to look past Bill to determine the reason. Looking back at Bill I was puzzled by the alarmed look in his eyes and his rather white face. Before I could say a word the shaken young man gasped "Christ, did you see him too?" I shook my head and asked what the hell he was talking about, whereupon he added "a bloody chap just walked straight through us!"

Notice in this instance I *felt something*, Bill *saw it !* One of us a student of the paranormal, the other a hardened sceptic.

This second shared incident occurred in the same drawing office. Ted R., Ph.D., technical author also had very strong sceptical views about the paranormal. While standing at my drawing board he had been reminiscing about a previous discussion we had in which he denied there was a case for considering paranormal effects. I had been in the process of stating that these things have to be experienced and one should not depend entirely on lip service. Almost as if in response to this, my other worldly helper tapped the usual three times on my right wrist which at that moment was resting on the drawing board. By now I was quite used to this phenomenon and the fact that when it occurred one could actually see the skin indented in just the same manner in which it would be if say a pencil was used. Feeling rather desperate to convey this as an appropriate example I described this phenomenon to Ted adding "You see what I mean, this is evidence to myself only, much as I would wish it I cannot share it with you" As I glanced up at him Ted seemed a little confused and - rather abruptly I thought - brought the conversation to a close and returned to his desk.

On several occasions during that afternoon I happened to glance round and spotted Ted pensively twiddling a pencil while looking in my direction. Hoping I hadn't offended him by the nature of our discussion, I shrugged the matter off until towards the evening Ted came across. He said he didn't want me to misconstrue what he had to say, but nonetheless in fairness he thought he should tell me that; "At the very moment you would have felt three taps on your wrist, I distinctly saw three flashes of blue light shoot down towards your hand. I have no idea what it was and I have never seen anything like it, but I thought I should tell you I *did* see this".

Another helpful corroborative visit

The following third instance which comes to mind occurred after we had moved to the Isle of Wight. I have chosen it not only because of the PK and corroborative content but as a simple example of the technically helpful nature of the phenomenon.

Although I was working at Saunders Roe at this time, as discussed in Chapter 12, I was also researching freelance for Lesney's Matchbox. This work was conducted in a spare room in our cottage and   the case in point   involved the rocket toy   shown in that chapter.

The launch pad for this toy was motorized so that it could trundle out from the central position to the outer perimeter of the toy. In order to achieve that I had borrowed several truck bogie wheels from my son David's 00 gauge model railway. To these I fitted electrical pick-up brushes between the onboard driving motor and the track. (in the normal model railway manner). To achieve realism I designed an indented flush track rather than a surface mounted one. This was formed by the simple expedient of sandwiching two strips of aluminium foil between several sheets of plastic thus leaving a small gap to locate the bogie wheel rims and contact the foil strips. Connectors were to be secured to the ends of the strips to take the electrical power input.

Having spent hours drawing the unit up, forming the sheets, fixing the rail 'track' and gluing prior to *impact* bonding, I suddenly received the now familiar three taps followed by an emphatic NO. I only had seconds to spare before the adhesive set, so responding with a hurried 'thank you' checked everything over, but the thing was so simple nothing could possibly be wrong; the sandwich sheet, the foil, holding down weights, *nothing.* So not for the first nor the last time in my life, I went ahead and did it my way. The sandwich fitted beautifully, the track width and depth perfect. The bogies sat square and ran smoothly, there was nothing amiss. Reassured I went ahead and built the rest of the track paving, fitted the motor and electrical pick up brushes and completed part of the central tower.

Now as any model engineer will tell you, it is not uncommon for small model railways or 'Scalextric' systems to 'play up' now and then. Dirty track, wheels and/or brush contacts can play havoc just when you don't want them to. Therefore I was disappointed but not unduly concerned when I put some power on and nothing happened. So, as always, I went straight to the track with a meter and found an open circuit. Sometime later I discovered that my recently purchased 'aluminium' foil, wasn't aluminium foil at all; I had been sold *electrically non-conducting coloured plastic!* O.K. so I should have checked it out, but that really isn't the point is it?

Carnival night and a synchronistic precognitive dream

In this same period I experienced other precognitive dreams, the following is one of them. In this I was standing with my wife in a crowded town square in Yarmouth on the Isle of Wight. It was a cool summer evening, the square was brightly lit and there was music, laughter and gaiety everywhere. Obviously it was a festivity of some kind. Suddenly I was violently jostled and quick as a flash a rapidly expanding gap in the crowd erupted. A brawl had broken out between two drunken lads beside us and I was doing my best to keep my wife away from the blows of flailing fists. In a moment the two were on the ground, one of them laid at my feet. The dream terminated with someone launching himself at the body on the ground and raining sickening kick after kick on the victim's head. Again I awoke in a cold sweat. This was exactly manifest in Yarmouth square the following weekend. It was Carnival night and my son Gary was playing in the band. But it wasn't so much the initial crowd jostling and even my recognising a large distinguished rather incongruously dark suited official that brought dream experience face to face with reality, but the instinctive reaction to protect my wife mingled with the horrific sound of boot against head.

It is interesting to note that only at the time of writing did the author realise there is a hidden element of synchronicity in this episode which linked it with the event included in Chapter 1. This would suggest that this sort of thing may occur in many people's lives which in the normal run of events are not recognised let alone recorded.

The relevant details of two events separated in time by twenty two years, set out here for comparison. Note in particular the *emotional content.*

1. In the *summer* of 1935 *I was 16 years of age.*

2. I attended a *festive occasion.*

3. I was going to play *my guitar and sing.*

4. This event was *spoilt* (for me) by my *falling with tangled foot.*

1. In the *summer* of 1957 *son Gary was also 16 years of age.*

2. I was attending another *festive occasion.*

3. I had gone there to hear *Gary play his guitar and sing.*

4. This event was also *spoilt* (again for me) by someone *falling at my feet.*

Finally I should point out that there was no parental influence on Gary's musical interest, for by the time he was one year old I didn't even own a guitar.

During the years in which I experienced the foregoing personalized phenomenon, there have been many occasions when I gratefully received prior notification of some important events (emphatically *not* of the state lottery kind). Such as an assurance to the contrary when, having awaited the arrival of a long overdue guest, all logic would have us believe the meeting had been unaccountably delayed or even incorrectly dated, when our guest duly arrived. Or a tap on the back 'cheer up' gesture at a stressful boardroom meeting. Or giving a lecture, a TV interview, at home or in the office, not least being prepared for a surgical operation. I have lived with this association with grateful acknowledgement and without abuse. The 'phenomenon' has been of an occasional nature as that of a friend, rather than overbearingly continuous. Moreover it has to be said there have been occasions when I chose to ignore it, but I also gave up trying to *prove* it ages ago. Indeed this will be the first time I have publicly disclosed this subject and for that matter to most of my family.

From the foregoing it might be inferred that on certain occasions, either the subconscious mind can function as an incredible all-seeing, all-knowing 'probe' or a discarnate and/or invisible (to our eyes) intelligent agency is independently involved. I feel the one proposition is no less intriguing than the other. However this kind of weighing up has been rendered less difficult for me by viewing these and many similar instances together, so that of the various possibilities I find the interpenetration of a plurality of worlds the most likely solution. Indeed I have been assured of such. That being so, then perhaps we shouldn't be surprised if - due to devoted practice - now and again we receive glimpses of what *for us* is to come, or for that matter what may have gone on some time, somewhere, before.

Neither is the author alone in this contention, for it was the well known tailless aircraft designer and mathematician J.W.Dunne who in the middle decades of the twentieth century experienced similar precognitive dreams to the present author, also involving his work in aviation. So much so that he made a study of his and other people's dreams which resulted in his writing his famous book titled *An Experiment with Time*. In this he proposed his *parallel universe* and *serial time* theory.

In this Dunne quoted some of his own dreams as well as those of members of his family and came to the conclusion that even people who claim not to have dreams at all, in fact do and further we all have precognitive dreams of some kind or another which are normally unrecognizable to us.

While in South Africa Dunne dreamt he was standing on a mountain in the middle of an island, 'watching smoke rising out of cracks in the rocks'. He became certain there was going to be a terrible volcanic catastrophe similar to the one on the island of Krakatoa twenty years before. In his dream he was pleading with the French authorities to send help and he woke up shouting, "Listen! Four thousand people will be killed unless..."

A short time afterwards, Mount Pelee on the island of Martinique did in fact erupt with the loss, according to a newspaper report, of 40,000 people. The newspaper figure was exaggerated - because of the similarity of the figures and the ease with which a nought could be left out - Dunne believed he had not 'witnessed' the catastrophe itself, but in his dream he had read the newspaper account.

An analysis and
a theory

There is so much evidential information concerning precognition, time-slips, synchronicity, bi-location etc., it is extraordinary that relatively so few people are actively impressed by it. This, I feel is largely due to the lack of a coherent theory. In other words the situation might be analogous to an artist having all the coloured paints in the world, but not having the slightest notion of the canvas on which to lay it. The following is an attempt to offer a little towards recognizing that particular canvas.

The dream experience is similar to the waking experience with the exception that the *normal* coherence to events appears to go by the board. For example, in a dream I find myself on a bus going to a certain destination. After an interval in the trip I suddenly find I am no longer on the bus, I'm now in an ambulance and hey presto the situation is changed. We cannot imagine - neither would we normally wish for - a real world state like that. The real world is different in that we *select* certain trains of events, whereas in the dream world we *appear not to*. For example, in the real world state suppose I had decided to travel on the bus. Having done so, suppose I fell asleep, during which time the bus was in collision and I was knocked unconscious. On waking I find that I am now travelling in an ambulance on a journey which I certainly did *not* select. Therefore it can be shown that the *continuity* of our so called *normal* waking experiences can in *certain circumstances* take on something akin to our memory of dream experiences. In other words there is nothing to prove that our dream world experiences are not just as real while continuities only *appear* to be disjointed and random due to similar unregistered interruptions. Indeed adepts in the art of total recall claim allegiance to this fact.

Thus it could be that in waking or sleeping the individual gains experience through which the soul - or 'overself' - develops as surely as plants grow in the soil. This experience can only be gained by pursuing differing environmental stimuli. Imagine a world where all the people were identical to yourself, where there was only one colour, where all flowers were of one variety, where we all had identical thoughts and agreed implicitly with one another, where all houses, even beautiful mansions, were all the same. We would have no reference point from which to look either *up* or *down*, indeed we would have no *experience* at all!

Therefore we *need* the profusion of variety of a planet from which to learn and develop. So we arrive here and experience differing stimuli by experiencing the environment from different points of view.

In order to provide us with the myriad of changing variety, the designer must have implanted certain programming so that people, for instance, would be different one from another. In humans - and for that matter some animals - this is expressed as conscious abhorrence to inter-breeding or incest. Even among flowers evidence of some degree of *programming* is evident.

Making sense out
of nonsense

There are some who find it more comfortable to believe that there is no order in the universe, that everything happens by random chance and that we do have free will. On the other hand, among those who are impressed by the evidence for precognition (or foresight) are those who find it difficult, if not impossible, to reconcile these two views. They quite fairly point out that precognition suggests that certain events are going to happen, which implies a kind of 'Fate', or that our lives are 'preordained', which of course negates the more acceptable notion of 'Free will'. In a word we cannot have both.

Such a thought can be abhorrent, for simply put it seems to suggest that we are little more than automatons, carrying out the will of our maker in perpetuity for reasons best known to Him. This paradox is born solely of our limited knowledge and it can be reconciled by the following hypothesis. That is, we do have free will and we exercise the prerogative by *choosing our lives on this planet before we came here.* Therefore once embarked upon, the wheels of 'fate' are set in motion and in that sense only we are *preordained* (or programmed) to fulfill our set mission. Beggar or king *we* choose it. But having chosen it, we have elected to accept the 'subliminal' programming and genetic inheritance our bodies might bring. Thus it is for some of us there is no mystery about certain facts of life that yesteryear seemed impossible to understand.

This becomes more acceptable if we consider that just as our subconscious minds tell our lungs to breathe in and out in order to replenish our physical body for growth, so the overself, ego or soul, likewise probably craves for all kinds of physical experience for development. Earthly experience is the *food* for the soul, and *we* choose it. Thus it is, that in this broader sense, fate and free will are not separate, conflicting tenets, indeed they are inseparable.

To a limited degree only, we also have the right to exercise free will while incarnate. The following more graphic analogy may suffice to place that in a hypothetical scenario.

A participant has elected to drive from his home at point A, in order to visit a friend who is in hospital at point B. There is only one road he can use and *within the scope* of his 'free will', only two choices he can take. He can either, 1 drive on the correct side of the road and travel *with* the traffic, or 2 drive on the wrong side of the road *against* it. In either event he will still end up *completing his set journey to the hospital* at point B, but his experiences on the way will be somewhat *modified,* i.e. he may go by ambulance.

The time college

To date there has been nothing to indicate there was anything of a precognitive nature about an experience I had in the winter of 1987, neither do I expect there ever will be. Although it was for my own satisfaction and therefore regretfully cannot be fully shared, nonetheless, it has provided me with added assurance that this sort of thing does occur to many of us.

I had already been acquainted with the hypothesis that at times, while the physical body sleeps, the ego , or soul, has additional OOB (out of body) experiences, some of which may be of an educational nature. Sometimes on waking the person remembers some of this experience. Indeed there are cases on record in which two or more people, who, while normally living far apart, have dreamed of attending a classroom in a school in which they met and discussed some of the subjects being taught. Later there has been cross reference by these various parties, either while speaking on the phone or actually unexpectedly meeting. Sometimes when they met at 'the school' it was a reunion of an old association which they had several years before. Again cross reference has established this fact.

No doubt due to my own dream experiences, I had retained an open mind about this possibility, but no more than that. That is until it happened to me! The 'subject matter' of the experience was by no means profound, it could even be described as mundane, and so far I have read little significance in it.

Wide-awake in another world

I was standing in a large workshop which was no different to most other workshops. The place seemed familiar as if I had dreamt of being there before, yet it certainly was no workshop I had ever visited in the 'awake' state. There was nothing 'ethereal' about the place and it was lit only by the light from normal sized windows.

In one corner there were several benches and machines and along one wall, under the windows, was a further long bench, at which the figure of a man sat working. He seemed familiar and as I approached he looked up expectantly. I suddenly realized it was my late father as he would have been in middle age. On the bench there were various tools and paraphernalia and he smilingly held an object in his hands which he set down for me to examine. It was simply a cone artifact measuring about five inches in diameter and seven or eight inches deep.

I picked the object up and found its weight was about one pound. In texture it resembled, and in fact I took it to be, glass fibre. Although fairly shiny, as though varnished, it was not very smoothly finished, almost as if it had been cast. It was brightly coloured, the lower half being royal blue and the top half a deep yellow. But the terminator between the two colours was not at right angles to the vertical axis, but oblique at, about forty five degrees, thus clearly defining the semi elliptical section. In fact I was able to separate the two halves which showed the flat section even clearer. Indeed the object reminded me of an exercise model in basic stereometry used at school.

During the time of this intent and fascinating scrutiny, I was aware of my father's quizzical amusement, however I felt sure there was nothing very special about the artifact itself. Suddenly, almost as if I had just woken up, I realized that this experience was just as real to me then as the reader knows at this moment he is reading this book. Perhaps at home or on a train, wherever, the experience was real, the memory of it is the unreal part!

It was with an extraordinary feeling of exhilaration that I looked at my father and said "I'm dreaming this aren't I ?" His smile broadened and seemingly amused by the situation he nodded his head in confirmation. It is most important that the reader tries to understand that this experience was so real, that as I opened my eyes and found myself in bed, it was something of a shock. I tumbled out of bed and relayed the strange story to an equally amused wife and as I told it, I knew that before the end of the day this real life glimpse of another world would begin to fade. But I resolved never to let myself forget that *this was real* and it *did* happen. I suggest the *substance of the experience* may have been of little consequence, but the *conscious identification* of it was.

From this it is logical to suppose that, as with many of us, I may have had other 'real' sleep state excursions about which I have no *conscious* recollection and this experience was merely to illustrate that. In which case it would seem there are no entrance fees for this college other than a keen desire to *know,* which probably induces a process akin to selective tuning and eventually a degree of further subliminal learning.

Neither is this course 'special', it is open to anyone and it is a short step from accepting this, to accepting the idea that the *overself* of many individuals assemble at places of learning, where during a few hours of deep sleep they accrue the equivalent of many hours or even days of extra-dimensional study. Such a process would explain how the dunce of yesterday, after being subject to certain stimuli (often as not a physical or emotional experience) might become a genius almost overnight. It could explain how it was by seemingly sheer *coincidence* in 1939 I was able to knock on the door of my friend Mr.C. who had coincidentally moved from my home town London to Trowbridge in Wiltshire, as discussed in Chapter 1. It offers an explanation for how it was that Jonathon Swift described the two satellites of Mars, Phobos and Demos before they were telescopically recorded. It could also explain how in *1898* Morgan Robertson was able to get so miraculously near to true facts in his novel *Futility,* when in this plot he described; A new unsinkable steamship of 75,000 tons called *Titan* that hits an iceberg in the Atlantic at a speed of 24 knots on her maiden voyage during April and sinks with great loss of life. The 66,000 ton *Titanic,* also claimed unsinkable, also on her maiden voyage, hit an iceberg in the Atlantic in 1912 at a speed of 23 knots. *Titan's* length was 800 feet against *Titanic's* 882.5 feet. The *Titanic's* twenty lifeboats (four fewer than *Titan's*) were quite insufficient for her 2207 souls on board, fifteen hundred of whom were drowned! The few cases listed in this book are representative, there are many, many more.

The origin of the teachers &
the potential of mankind

The question might be raised, what is the origin of the dream world teachers? For instance they might represent souls from a higher hierarchy whose roots stem from all over the physical and metaphysical universe, much the same as we have multi-national teachers in our waking colleges on earth. Some of these souls might elect to incarnate here in order to continue their work and thereby gain further experiences for themselves in the process. Some of them might be souls who lived on earth centuries ago. Yet others might originate from different planets at the physical level, then freed from the restrictions imposed by distance, time and space continue to function at the metaphysical level of their planets, and others originating at interdimensional levels. There is no limit to the strata of life that we can imagine. This should not be surprising when we consider the myriad of varying organisms which exist on our own planet alone.

Predictably therefore the skills of these masters are beyond our comprehension and on brushing shoulders with them now and then, we shouldn't be surprised by their mastery over matter. In fact we are only just beginning to have the barest idea of what this can mean by accepting the gifted talents of people like Uri Geller. Make no mistake, young children have, and do, bend metal objects by other means than by force. In this respect it is difficult for some of us to realize that as far as technology and a *true* understanding of nature is concerned, we really are just beginning. A somewhat inadequate analogy of the situation is to remind ourselves that the primary reasons for one of the most prevalent physical ailments which bedevil human kind at this present time, that of back trouble, is due to the fact that in purely physical development rate times (going back millions of years), homo sapiens only began walking upright on two legs *yesterday !* Moreover it is generally accepted that at present we use less than twenty five per cent of our brain. Thus, bearing in mind the exponential rates of development I introduced in The Cosmic Matrix, the final capability of the human brain must obviously be beyond our present comprehension.

Finally at the risk of straying slightly out of context at this juncture, I am reminded of the many discussions I have had with people concerning some of the more sombre implications of the foregoing, not least the implied vexing question regarding reincarnation of the personality or soul. How can it be, some have asked, that given the choice some would freely elect to be born to a short life of abject misery and suffering as a little child? To which question I would respectfully answer by posing another. What teacher would give his best and most advanced students an easy task? Indeed, rather than reflecting punishment for past wrong deeds - as some misguided would-be religious scholars are inclined to have us believe - for all we know, such suffering for some young minds might be a self determined experience which few of us would care to contemplate. Moreover wasn't it Jesus of Nazareth, who when asked by his disciples in what had this child or his parents sinned that he should be born blind, answered `Neither but that the works of my father which is in Heaven should be manifest unto him'. In addition to which I hasten to point out, if - as I have suspected all my life - our world is a kind of cosmic school, then it would be pointless for it to be populated by people of one class; all saintly souls for example. As I have stressed elsewhere, there has to be contrasting elements among society as there are contrasting colours in a garden from which we derive experience. I also suggest that this premise is not restricted to humans and it is a grave mistake to believe that *all* Ufonauts, for example, should be painted with the same brush.

The latter portion of this book has been devoted to some results of my `ordinary' precognitive excursions, (typically of the nuts and bolts kind), and I am all too aware of the fact that they pale into insignificance when compared with the works of others. Equally however, I must reiterate that although not as profound as all that, what precognitive tendency there has been revealed here should not be neglected from the major findings in the Cosmic Matrix. I trust this point will take on an added significance by the inclusion of some of the events revealed for the very first time in the following chapter.

PART TWO

The UFO connection.

3
Angels Unawares?

1953 - 1961

From the outset of this section I must stress that while most of the following incidents occurred a long time ago, they were faithfully recorded and can be verified. Also although I am unable to meaningfully correlate them with some of the precognitive episodes in the rest of the book, beyond having fleeting suspicions now and then, I am content to leave that to the discerning reader.

During the years, with the exception of several people close to me, I have never before publicly divulged most of the information in this chapter and even now having no wish to cause inconvenience or embarrassment to those concerned they must remain anonymous. In setting out the introduction to what follows I am mindful of the need to quantify a present ambiguity which still exists among some prominent sections of the UFO community. In particular, I refer to the apparently perplexing fact that among the various UFO occupants visiting our planet there has always been humanoid and even perfectly normal human beings breathing our atmosphere.

As a long standing Ufologist and aerospace engineer I too had to reconcile this apparent anomaly many years ago, so with the benefit of hindsight the solution now seems obvious enough. However in the past the problem was severe enough to virtually split and disrupt UFO groups right through the middle the world over. Neither was it only the groups, I have known many Ufologists to remain at loggerheads because of it. However the problem immediately vanishes if the prevailing status quo is reversed, in which we imagine we had developed our transport and space flight technology to the standard of that set out in my previous books, in particular The Cosmic Matrix. Journeys to our neighbouring planets and deep space have been achieved and exploration has been in progress for some time. Say, perhaps within our solar system - on a satellite X for example - we had discovered another indigenous species of relatively undeveloped cultural status. Not significantly different from us physically, but adapted to different environmental conditions unsuitable to ourselves. Given that we had an important enough need to be there, we could only sustain activity with space suit, breathing apparatus etc. Again, given the importance of our presence on the satellite it would be logical to give the resident species a helping hand. Provided that need was sufficiently dire it would be benign and mutually beneficial for us to nurture and educate some of the indigenous offspring among which eventually would be teachers, scientists and space flight crews, who could if they so wished be returned to their place of origin Satellite X to work, help and perhaps later even re-cohabit in an entirely clandestine capacity.

The whole point being, over the years some of the denizens of Satellite X would be witnesses to strange aerial contraptions and wonders from the skies. From these would come we, the *alien* looking occupants in space suits and now and again 'people' just like them breathing their atmosphere. In terms of this analogy, if the planet earth has been visited by aliens for a considerable time, then it would be *strange in the extreme* if among them there were *not* quasi human pilots, scientists and even some tourists returning to *their* planet of origin once in a while!

However, bearing in mind what has been discussed in the previous chapters and in The Cosmic Matrix, there is an additional possible explanation for human occupants of some UFOs. It is that of dimensional transference, or time travel. Throughout my research over years I have been obliged to take note of the fact that among the many UFO contact reports, which generally might otherwise have stood up to comfortable debate, there have been those which have been smirched and relegated to the trash bin primarily due to the fact that the planet Venus has been indicated as being a home base. I suggest that although understandable, eventually this may prove to be erroneous, for in addition to the foregoing scenarios, it so happens that *for a technologically advanced* race, the engineering and covert requirements may be less formidable on Venus than they might be, say, on the Moon.

The Adamski/Coniston
photos analysis

In 1953 Desmond Leslie and George Adamski's book *Flying Saucers have Landed* was published. In February 1954, a thirteen year old boy Stephen Darbishire and his eight year old cousin Adrian Myer photographed a flying saucer at Coniston, Lancashire. Note, at that time he had a persistent restless feeling that he had to go there. Despite the fact that the resulting photos were out of focus and at a different angle, the similarity to those taken by George Adamski was obvious. Moreover there was sufficient dimensional detail to enable me to do the now well known Adamski Coniston orthographic analysis first shown in my book *Space, Gravity and the Flying Saucer, 1954.*

Waveney Girvan editor of the publishers T. Werner Laurie invited me to a crowded public meeting in London, where I met Desmond Leslie and the two boys. Later I was able to talk privately with them and was immediately impressed by their openness and sincerity, it was quite obvious to me they had nothing to hide. Of no less importance to me is the fact that although I had the technical know how for this investigative work, it was my paranormal and UFO sceptical friend Ted R, we met in Chapter 2, who recommended its instigation.

In this context I welcome the opportunity to rectify an inaccuracy which has been unknowingly adopted by some prominent Ufologists. It is the fact that they have relied heavily on the group of *three* portholes seen in the Adamski scout photo. In point of fact Waveney Girvan told the author that there had originally been a set of four, but he had to trim one out in adjusting the photo for the frontispiece format of Flying Saucers have Landed.

Shortly after the above meeting, two other incidents of a precognitive nature occurred. In the first I was invited by a friend Lt. Col. Donald I, to attend a private seance with a very well known London medium, some of whose patrons are known to come from very high stations in society. Apparently he had a very important message for myself..

The medium, a slight man in his forties, greeted us cordially and after a brief introduction asked us to be seated. After a short interval the medium became entranced. Then in a different male voice he said I should try to accept this meeting was genuine and

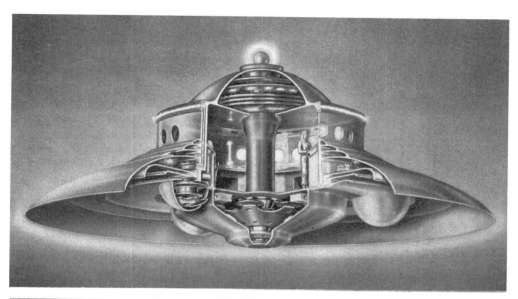

*The author's frontispiece and dust jacket for Space,Gravity and the Flying Saucer published 1954. The very first copy 'hot off the press' was hand delivered to Buckingham Palace followed by a repeat request within the hour. Having got the book serialised in several American newspapers, why did the British Book Centre in New York go into liquidation causing the almost total loss of sales, then mysteriously go back in business under the **same name** less than a year later? Whatever the cause, in Chapter 12 echoes of synchronisity are encountered involving another of the author's projects, another American company and the copy book mysterious circumstances associated with it.*

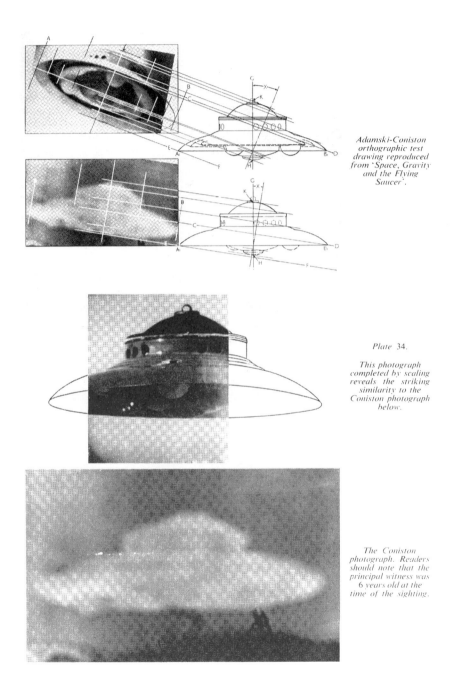

Adamski-Coniston orthographic test drawing reproduced from 'Space, Gravity and the Flying Saucer'.

Plate 34.

This photograph completed by scaling reveals the striking similarity to the Coniston photograph below.

The Coniston photograph. Readers should note that the principal witness was 6 years old at the time of the sighting.

The Adamski/Coniston orthographic analysis drawing is reproduced from Space, Gravity and the Flying Saucer and Piece for a Jigsaw. Did some factions find this simple exercise a little too inconvenient?

sincere. After a few words this 'control' said he came to introduce an individual who wished to impart knowledge to me and for that specific reason the meeting had been arranged. There followed a few moments of silence before yet a different male voice spoke softly and hesitantly. First apologizing for difficulty in using 'this technique' then the voice gradually became more audible and stronger until the control was able to describe himself as an individual whose origin was not of the planet Earth. How often I had heard it all before! As if to reassure me, I was given a very accurate description of some of my activities during the previous week, which I am sure was hardly likely to be generally known. This culminated in the voice telling me of my movements that very day, even to how I had popped into Woolworths Store to purchase a blue covered school exercise book for more notes on gravity and space!

I was also told that little book would be important to me one day and that I had a son - who was 10 years old at that time - who in years to come would help me in my work. As will be seen later on, all of this *did* happen. To this day my son Gary *is* a great help to me, as is his brother David born after that time. I still have that little blue book, now time honoured and worn, but still as valuable to me as it was then.

But now I was rather taken aback by the nature of the next 'prediction' when the voice said that very shortly I was going to have a very important meeting in which a space craft would land and I would be met by people from another place. I still remember the last words concerning this were, "You will be gathered up into the bowels of this *saucer*".

The rest of the discussion concerned both sociological and technological matters and not least it was a very impressive endorsement to the revelations which were coming my way. I had an odd feeling that the owner of that strangely familiar voice out of the darkness and I, had met before and would meet again.

A precognitive dream
of two events

The second incident was another vivid precognitive dream, but this one was directly UFO related. In this I was standing on the open deck of a ship at sea looking towards the destination which I could see as a distant familiar shore line. Then suddenly I was standing alone in meadow land with my back towards a house and the land rising slightly toward some trees which also bordered one side of the area. Despite the fact that it was nearly dusk, I could plainly see an exact replica of the Adamski scout ship gently descending about fifty yards ahead of me. When it was only two feet or so above the surface, it unexpectedly dropped making a slight audible thud, which I could almost feel through my feet. In the dream this both surprised and concerned me.

The next moment two figures stood before me who somehow I *knew* instantly. One was a tall muscular male, the other a shorter female. They were deeply tanned, had jet black hair and were dressed in what reminded me of typical North American Indian style, complete with bare arms, beautifully decorated head bands and painted facial markings!

In the dream this did not appear strange to me, I knew them both and with great exhilaration - I find it difficult to describe - I hurled myself towards them. Indeed at that moment I felt and acted just like a child. On reaching the man I flung my arms around him and unashamedly sobbed with joy, I could even feel tears running between my face and his bare chest. He clasped me with great warmth as a parent might and looked meaningfully down at the lady I felt was his wife. As discussed in the last chapter, allowing for the fact that in dreams extraordinary things happen, what happened next I still find hard to set down. I knew they were conversing and understood the gist of their conversation though no words were spoken. Then with a steady compassionate gaze the man gripped my shoulders with one hand and taking hold of my hair (which seemed to be long) with his free hand he pulled my head backwards. The motion was both deliberate and quick, the look in his eyes unchanged. Then without warning he released my hair and with the edge of his hand *struck a sharp 'Karate chop'* style blow just above the bridge of my nose. This both shocked and hurt a bit, but I was instantly reassured by the look on his face. It was a sort of 'sorry about this' but it is necessary look. I knew the two were still conversing and there was a feeling of anticipation about whether 'it' was going to work. After a brief pause he struck again, three times in all and each time with that look of concentrated hopefulness. The second and third times didn't hurt so much, more like a numbed smarting than anything. But stupidly in the dream I was anxious he didn't break the glasses I normally had to wear. Here I woke up with mixed feelings, wondering how on earth my imaginative mind could conjure up an emotional scene like that!

In this respect the reader will notice an identity with the childhood garden experience of Uri Geller as quoted in his book My Story, and as we shall see several *years later* I *would* be standing on the deck of the ferry travelling to our new home on the Isle of Wight, where the meadow on which the scout ship in the dream appeared to land, would turn out to be the grounds of that new home! (see photo)

The 'space indians' part of the dream is hardly ever likely to manifest. Occultists will claim that it is purely 'symbolic' and I have little doubt some psychoanalysts would have no difficulty explaining it. That being so it would be interesting to have their explanation for the proven precognitive remainder which follows later on.

Space,Gravity & the Flying Saucer is published.
Then mysteriously withdrawn

In 1954 I was amending the manuscript of my book *Space, Gravity and the Flying Saucer* to include the Coniston-Adamski orthographic analysis drawing, but in the midst of that, yet another helping hand was offered, which turned out to be the most extraordinary of them all. In the book I had tentatively portrayed a somewhat mechanistic analogy of gravity in order to give the lay reader an idea of how *etheric* flows to and from a planet might occur. But my author friend Ted R. pointed out that an Englishman, Antony Avenel had already published a much advanced theory which was far more embracing than my more gravitational aspect. As this work was totally independent, Ted R. suggested it could lend support to my own findings. Why not contact Antony Avenel? This I did and was amazed by the most generous response. Not only would Avenel allow me to quote his own independent conclusions, but he offered me the free use of his entire work! The title of this concept was of course *The Unity of Creation Theory* and the moment I looked at it I had that familiar sense of déjà vu. As is now known, I was privileged to take advantage of this remarkably generous offer and the abridged version of the theory was duly incorporated in the original *Space,Gravity and the Flying Saucer* manuscript, and indeed my subsequent books.

I still have memories of the bright sunny day in May 1954 when I had just arrived at Werner Laurie's publishing house in London to discuss the launch of the book with Waveney Girvan when Derek Dempster, then Air Correspondent to the Sunday Express, (whom I had not previously met) swept breathlessly into the office to state that by special request he had just delivered the first copy of the book 'hot off the press' to Buckingham Palace, and that he had now been asked to deliver another copy.

The little book was subsequently serialised in several American newspapers, being the third ever published on the subject and the *first with a corroborative scientific appeal.* Yet before a mere handful of the books could be sold,the British Book Publishing Centre of New

York went into voluntary liquidation. A check on the situation a few years later revealed they were back in business *trading under the same name.*

A space oddity?

In March 1954 I received a letter from Patrick Moore inviting me to join him and Desmond Leslie at their club. I was seated beside Patrick and over dinner, despite his claimed scepticism about UFOs, we had a very pleasant evening. However I was surprised to learn that only a week before he had met Cedric Allingham, author of *Flying Saucer from Mars* and he was keen to discuss this book with me. Apparently in this Allingham had made some observations on the similarities between the Adamski and Coniston photographs and those he allegedly took of a grounded scout ship at Lossiemouth, Scotland. In particular he had also made some reference to Patrick about my orthographic projection tests which made it quite clear to me that Allingham was unacquainted with the technical process involved and therefore on these grounds alone the comments attributed to him were invalid. Moreover in so far as Allingham had made an issue of the number of portholes in the saucer he photographed i.e. *three* he was no doubt erroneously influenced by Waveney Girvan's *trimmed* scout photograph as mentioned earlier.

At that time the extraordinary thing was, that other than Patrick Moore, nobody else had met Allingham and all subsequent attempts to locate him failed. It was said he had been very ill, but where did that information originate? It could hardly explain his total disappearance. It was generally suspected that *Allingham* was a pseudonym, for it was known he had written S.F. novels under another name.

Now I had examined the photographs from Allingham's book very carefully and arrived at the conclusion that they were not very good fakes, and a small model had been used. In the shot showing the alleged landed scout craft, part of the rim and one spherical 'undercarriage' was completely out of focus. While a slender 'antenna' in the middle of the dome and the farthest spherical undercarriage were quite sharp. It is most unlikely that such distortion could be explained by the localised gravitational field optical effect I was to predict in my second book *Piece for a Jigsaw* in 1954. Moreover Allingham had also claimed his saucer was grounded and inoperative.

But of more importance here is the fact that according to my informative source, this was a very clumsy exercise perpetrated by silly people who had other interests at heart. Who unknowingly were used to disrupt the technically corroborative evidence for both the Adamski and Coniston photographs which that part of my book *Space, Gravity & the Flying Saucer,* illustrated in some useful depth.

The first UFO landing
near our home

By the summer of 1954 UFOs were drawing closer to us and the reader is asked to bear in mind the forestated medium's message in which my son Gary was specifically mentioned.

At that time my family and I were living in a small village called Bricket Wood near St.Albans, Hertfordshire,UK. Gary occupied a bedroom on the first floor at the front of the house, and my wife and I occupied an adjacent room. At approximately four o'clock one morning I was abruptly awakened by Gary standing at the foot of our bed, as I thought bathed in moon light, urging me to rise and see something odd. He was very excited and ran to the balcony window. Joining him I could see nothing, but he insisted we hurry into his room for a better vantage point. We did so, and with a 'there it goes' he pointed towards a nearby oak tree,and I was just in time to catch a glimpse of a bright light source disappearing behind the tree, then there was a bright flash which accentuated the following darkness.

He then told me his story. He said "I suddenly found myself awake, I don't know why, looking at the moon shining through the curtains at the window. Then I found myself thinking, the moon;'s not that shape and it shouldn't be moving like that and this woke me up properly. It was so bright I could see it shining through the curtains as it slowly rose higher and higher, rocking a little from side to side". Going to the window Gary then saw the UFO more clearly. "It was glowing white and fuzzy looking and shaped something like a banana. Then it stopped rocking, tipped over and started to slip downwards. Then silvery dust seemed to sparkle and drift away from it. Then I called you".

The following morning I called the editor of a leading national newspaper in case others had seen the UFO. He advised me to tell my son 'to take more water with it'. Gary was thirteen at the time!

Now it would be natural for the reader to believe that, to some extent, Gary had been influenced by my interest in the subject. But in fact the boy was more interested in natural science, almost to the exclusion of other topics including space travel and saucers.

Reconstructing the event, I was able to take measurements from the position of the bed and through the window. A compass bearing and a map placed the line of sight directly *through the middle* of the Handley Page Airfield at Radlett three quarters of a mile away. Gary's rough estimate of distances placed the object somewhere near the runway. From this I calculated the altitude and estimated the angle of descent, and again referring to the map found a place of intersection on the ground. This turned out to be nearby heathland. We visited the spot and found a large area of burned bracken which looked fairly recent. I was unable to check the cause of this.

I had almost forgotten this event when, nearly two months later, two colleagues of mine, (both engineers working at Radlett aerodrome) visited me and somewhat furtively told me of a 'leak' that had gone around their firm. It appeared that on the *same morning* Gary stood at the foot of my bed, the night staff at Radlett aerodrome had seen a large glowing thing take off vertically from the runway. The whole thing had been hushed up and I was asked to be discreet. The reader will imagine the looks on my friends faces when I told them the sequel!

View from Orbit Two

During this time the generosity of Antony Avenel was repeated and for the first time I have yielded to the wish to record the following extraordinary events. Inevitably, rightly or wrongly, readers will no doubt form their own impressions. But most importantly, it is my sincere wish that the following information will in no way inconvenience anyone concerned, and as in the past, I shall always respect individuals anonymity.

During my correspondence with Antony Avenel, it became clear to me that there was indeed infinitely more to the Unity of Creation Theory than the mere gravitational aspect which had first attracted my attention. In one letter I had suggested that many people would be keen to know more about the theory, why not a book? To which Antony said he would consider it. Shortly he wrote saying he had decided to publish the theory and as he was so busy with other work, he asked if I would be prepared to help, perhaps with the dust jacket design etc. Of course I was delighted to help in any way I could, this at least, I thought I owed him.

Until then I had little idea what this entailed , so I was intrigued to receive Antony's basic proposal in which the Earth and its present culture would be reviewed in terms of the Unity of Creation Theory.This to be presented as observed from a detached independent source situated within the solar system. For this Antony proposed a wrap around dust jacket design depicting a perspective view of the solar system with elliptical shaped planetary orbits around the central Sun. The view point was centered on the planet Venus placed in the foreground, so I wasn't surprised when he entitled the book 'View from Orbit Two'.

Later, Antony asked if I would like to make an additional paid contribution, with the help of other writers, to cover some of the main issues presented in the book, while his contribution would be the conclusions on each of these aspects as embraced by the general theory. In a word the book migh be interpreted as a *message* from Orbit 2. (this somewhat speculative latter observation is my own and has never before been publicly stated).

Now it should be understood, that at that time I had a full time job on my hands, which frequently called for overtime. Any *one* of the dozens of items Antony cited in the text would normally involve many hours research at public libraries. Which meant I would have to research things and make voluminous notes before my wife could edit and type them, and the agreed publisher was waiting for the manuscript. The work alone could have taken months and frankly I was extremely troubled about my general ability to handle the task. But somehow I knew I could not let this man down, after all I had suggested the idea in the first place. Therefore I did not even hint of my dilemma. My technical author friend Ted, as always, stepped forward to help and meanwhile Antony had already arranged for another writer to contribute, which was a great help, but that still left us with a massive share.

Then during this time Antony said his cousin Brian would like to pay us a visit bringing the agreed rough dust jacket design and other relevant material with him. On 18th December 1954 Brian duly arrived, it was to prove another extraordinary experience for me. He appeared to be in his early thirties, was quite handsome, fairly tall and well built. After mutual introductions he excused himself saying that he had to fetch several books from his car. Irene and I were astonished when he returned to the house with a pile of books tucked under his chin with arms full length and smiling asked if he might put them on the settee. Seemingly enjoying himself he then returned to the car for another lot , and another after that, until by the time he had finished the settee was packed full of books! Joining in the fun of the moment, my wife and I selected one or two random copies to peruse. We were amazed to discover that none of the books were old or worn, certainly many of them were very expensive and brand new. Moreover carefully written on each dust jacket with page numbers and appropriate paragraphs was all the information I would require...my impossible extensive research had already been attended to for me! If my words were found wanting, I do hope the look of relief and gratitude that must have shown on our faces may have made up for it.

We had a meal and settled down to talk, Brian on one side of the fireplace and myself on the other. I asked if he objected to me smoking a pipe whereupon he said no, in fact he would join me. Now of course there can be little doubt that it may merely have been a coincidence, but to me his pipe was *brand new, as was his packet of tobacco* and we both acknowledged good humouredly, his light hearted embarrassment as he proceeded to go through the process of pipe filling and smoking, which to me as a pipe smoker , he had never done before in his life!

We discussed several outstanding issues concerning the forthcoming book and Brian asked if there were any further questions I had in mind. I looked at him for a moment then said "Why did Antony choose *Venus* as the vantage point from which to view the Earth?"Returning my look he smiled and said " I suppose he thought it just as good a place as anywhere else in the solar system." Then later that day, just before leaving, almost as though an after thought, Brian handed me Antony's updated dust jacket design. There were no alterations to it with the exception of the completed title for the book, It read 'View from Orbit 2' by Leonard G.Cramp! This took me completely by surprise and frankly I was astonished and not a little bewildered. Brian was sitting there nonchalantly awaiting my response, which when it finally crystallized was only a half stammered..."..but"...Obviously for some reason this gesture was Antony's wish, but by then I had been looking forward to *his* work being published under *his* name. I apologized to Brian and hoped Antony would understand if I declined this one further generous offer. To which Brian said he fully understood and would discuss the matter with Antony when he returned. Sometime later I heard from Antony saying he quite understood my position and he would sort it out.

Eventually, as 'View from Orbit 2' was nearing completion, Antony wrote to me saying that Brian had decided to accept a teaching job in South Africa and he was shortly travelling to London to sort out several matters and attend a dental appointment. He said Brian would welcome the opportunity to visit us again on route when he would collect Antony's books for him.

When Brian eventually arrived it was getting late, so had little time for discussion and in order for him to make a very early morning start , we helped him load Antony's books into his car. I still have three vivid memories of that last visit. The first, which despite the late hour, was when Brian had taken from his briefcase the revamped dust jacket and smiling handed it to me. It was still the same as when I had last looked at it , save for one modestly small addition. The title now read *Avenel's* 'View from Orbit 2' by Leonard G.Cramp! True to his word Antony had compromisingly sorted that one out for me.

The second memory stems from a moment when Brian had bid us goodnight and left the living room. A short time after I had quietly followed him into the hall expecting him to be upstairs by then. However he was standing at the foot of the stairs with our dear little newly acquired kitten in the palm of his raised hand, gazing into its large eyes in rapt admiration, murmuring softly as he stroked its furry little head. Extraordinarily enough, for an instant I felt as if I was unintentionally intruding or interrupting something rare or special. Whereupon Brian smiled approvingly, replaced the kitten on to the hall floor, said goodnight again and carried on upstairs.

Despite our friend's assurance that he could see himself off in the early morning, Irene and I insisted on being up to see him off. However it didn't turn out that way. For although we hadn't overslept, we awoke to find our visitor had already left. We were touched by his not wishing to disturb us and my mounting sense of awareness and intrigue was not lessened when I later discovered, not easily overlooked, splashes of blood in the bathroom basin. I knew that Brian had attended a prior dental appointment in London, but I would have sworn that he had a perfect set of teeth. I was concerned, but also began to feel as though a message had been delivered and now understood. To this day we have never heard from the one who called himself Brian P. Neither did I ever discuss this occasion with Antony Avenel. Somehow I felt I was not intended to.

After some initial problems View from Orbit 2 was finally published, but not before Waveney Girvan had insisted that the by Leonard G.Cramp part of the title was deleted. At the time Antony and I thought that Girvan was rather exceeding his mandate. However I was more than content to be given pride of place in the list of acknowledgements which included the name of a very prominent scientist.

I reiterate, throughout this work I have no wish to knowingly betray a trust, compromise, nor cause embarrassment to anyone concerned. I am merely placing on record some true accounts of events as I have found them.

To the best of my knowledge, to this day I have neither met nor spoken with 'Antony Avenel'. As also I regret the fact that *View from Orbit 2* - no doubt long before its time - did bot sell very well and was never reissued. Not least I regret having been foolish enough to entrust my only signed copy to someone else several years ago.

A prophecy fulfilled

In 1955 Waveney Girvan phoned to ask me if I could visit him at the London office. Arriving there the following morning it was immediately apparent he had a problem on his mind. He handed me a typed manuscript and said it was George Adamski's new book and he would be most grateful if I could spare the time to take the manuscript into an adjoining office and browse through it uninterrupted and give my candid technical appraisement of its contents.

I must confess I was surprised and excited to learn what Adamski had to say. Thumbing through the pages, one of the first chapters I came across entailed the description of the interior of the scout. Naturally I warmed to the fact that my own illustrated portrayal was quite close to his description. Although the book fascinated me I had reservations about certain aspects. Even so, it should be published.

After about an hour I joined Waveney and offered my opinion, but it soon became apparent that from a lay persons point of view, what was concerning him was the sheer mind boggling aspect of a gigantic space carrier hundreds of feet long and housing as many personnel. In effect Waveney was asking me if such a thing was technically possible. Of course I didn't hesitate to affirm this, adding that from a *space craft technology* point of view I could not disqualify the book. This seemed to reassure him and I duly left the office. However a few days later he phoned to tell me that he was seriously concerned about the validity of Adamski's claims to a point where he had decided against publishing the book.

However in 1956 Adamski's controversial book *Inside the Space Ships* was published in the UK by Arco Spearman, and as is now well known, it had the most dramatic splintering effects on the flying saucer community throughout the world. Many people I knew personally were at loggerheads over the claims he made. Then in a letter to me Adamski congratulated me on my artist's impression depicted on the dust jacket of my book *Space, Gravity & the Flying Saucer*. He wanted to know if I had ever been invited on board a scout ship or knew anyone who had, as he thought the representation was uncommonly accurate. He made particular reference to the central column of the craft shown in his sketch in *Inside the Space Ships*. His quoted remarks included 'It's the closest thing that anyone could possibly get'

In March 1956 my family and I took up permanent residence at Gardeners Cottage on the Isle of Wight. Thus the significance of the 'familiar' sea trip in my previously described dream was realized, as also was the meadow in which I stood, at the rear of our cottage.

The visiting generals

In April of that same year I joined the aerospace and test tank establishment then known as *Saunders Roe Ltd.,* at Cowes on the island. This company was eventually to become the world famous *British Hovercraft Corporation.* At that time the company projects included rocket and jet propelled (mixed unit) prototype aircraft designated P53 and P177 together with the Black Knight and Blue Streak rocket launcher programs. By that time my book had long been published and withdrawn, I had written articles, given lectures and appeared on TV several times. I was of course aware of great overseas interest being directed towards our new mixed unit fighter projects, visiting VIPs from other countries was almost the norm. Then on one notable occasion our then assistant chief aerodynamicist, the late Dick Stanton Jones, asked me to drop in for a chat, as I thought concerning another of my projects. He didn't beat about the bush but came right to the point. Had I published any work on gravitation in the United States before joining the company? As this was so, I answered in the affirmative. A broad smile of relief lit up his face and after thanking me he related the following extraordinary story.

It appeared that at the last meeting with several visiting Four Star Generals from the Pentagon - who had declared interest in our aircraft projects - both Dick and his chief were casually it seemed, paired off into their own separate parties. Dick said that he soon detected a strong feeling that the visitors seemed somewhat distant and were not quite so intently interested in the aircraft as one would have thought, rather they were asking random general 'fringe' questions with more than a hint or two about gravity. Dick and his chief were able to arrange a moment to confer and soon, realizing they were both getting similar treatment, decided to 'play along' with the situation. After several hours of this game of call my bluff, the charade came to an explosive and abrupt end over lunch, when the general sitting opposite Dick, while still in mid sentence concerning the aircraft, suddenly leaned forward and red in the face and with half strangled hiss said, "Damn it man, give us a break, what are you getting for your negative mass ratios?" Dick could only guess what the name of the game was and taking a stab in the dark proffered the first mischievous retort that came into his head. Apparently he acted his part suitably convincingly, and after furtively looking over his shoulder, he leaned across the table and whispered "about 0.05 or thereabouts". Whereupon after a cliff hanger pause the general beamed gratefully, slumped back into his chair and then as if nothing untoward had happened, resumed sudden rekindled interest in the aircraft!

For some time after, Dick and his boss were at a loss to know what the affair was all about, save for the fact that the American visitors were extremely interested in matters gravitational. Dick's conclusion was to me a little embarrassing, as after all I was a comparatively 'new boy' at the firm. He figured; the Americans on becoming aware of my gravity research work, had quite erroneously assumed that it reflected the interests of my employers, who in fact only knew a little of these private endeavours! This sort of thing not only happens in novels, TV and films, it happens in the real live industrial world and it happened to me. In which case perhaps I will be forgiven if I add that I have now and again allowed myself the light hearted, not very serious thought, as to how much the world's future may have hung on Dick Jones chance retort! But now, again in context of what is to follow, the question has to be raised 'Were our visitors from the Pentagon *solely* interested in gravity research status abroad, or were their motivations far more closely orientated towards the main subject of this book?'

A mother ship paves the
way for the future

Shortly after this a colleague told me of an interesting UFO sighting which had been witnessed by a Mr. Jarvis, the Saunders Roe boiler man. As he had since retired I called to see him. He proved to be a delightful old chap of local farming stock. He seemed to be quite healthy and articulate and had no objection to discussing the experience. If anything he seemed eager to have an explanation for it. Apparently it occured some time around 1954, before our move to the island.

He lived only a few hundred yards from the main offices and it was his job to arrive early to tend the boilers. It was still dark at 4.50 a.m. that particular morning as he walked along the lane that led to the main gates, when he became aware of the fact that he could see his shadow thrown on the ground in front of him. No sooner had he realized this when the whole area lit up with a pale green light. He could see the lane, the hedges on either side and the one or two bungalows at that spot.

Instinctively he had looked up and behind him and there, now almost directly over his head, he saw a huge 'zeppelin' shaped craft. It was moving, but not too rapidly for him to notice something 'like lit up portholes along the side' and a diffused kind of red flame at the rear. There was a total absence of noise. He also said it appeared to have a halo surrounding it and the object itself was a dark silhouette

Thinking it was some kind of 'secret hovercraft' he continued on his way until he met the night watchman going home. The two met with the latter exclaiming. "Did you see that bloody enormous thing?" I asked Mr.Jarvis if he could describe how big it looked to him and glancing around at one of the nearby bungalows he said "About the length that roof looks from here. I checked this out and it represented more than a hand span at arms length. Clearly he had seen something like the Adamski cigar craft. So fishing in my briefcase for my collection of UFO photographs, I selectively thumbed through them. When I came to the Adamski 'mother ship' his reaction was excited and spontaneous. *That* was what he had seen, "But for God's sake what *was* it?" he asked. I told him I really did not know for sure, though I had my suspicions.

The route of that craft was entirely consistent with a series of similar sightings that had run across the south coast of England during that time. I knew because I had already interviewed some of the witnesses before moving to the island. It seems Mr. Jarvis may have seen the tail end of the route taken by the craft as it crossed the island towards France, where it was also tracked. The name of the place over which the object described by Mr. Jarvis flew, was *Whippingham* .....at the test tanks of Saunders Roe! We shall be hearing much more about Whippingham later on.

This corroborative shared sighting of an Adamski type carrier was for me very gratifying and it was a great boost for my faith in him. As previously implied, if you are an engineer worth your salt, then you shouldn't let bias creep into a technical evaluation. For engineers to behave otherwise would mean that cars wouldn't work, bridges, skyscrapers and aeroplanes would fall out of the sky en masse. I think it fair to say this approach is repeated in my orthographic projection analysis of the Coniston-Adamski photographs, for had it proved negative I would have been the first to have admitted as much. My more recent technological findings with regard to the Adamski scout (shown in the Cosmic Matrix)is a furtherance of that stance. In other words I am obliged to accept the Adamski *photographs* are genuine. However although I would prefer to say as much for his claims, unfortunately I cannot. But equally neither am I in a position to apportion *blame* in this respect. The following may help to quantify my position.

So who was kidding who?

I had been invited to attend a meeting at Bournemouth during which I had the pleasure to meet Adamski which proved to be a memorable occasion. I spent several hours with him and although this meeting was all too brief, nevertheless I had time to ascertain that he fitted exactly the opinion of others who had got to know him. He was an extremely pleasant man with a good sense of humour which now and again was tempered by a feeling of deep sincerity.

Accompanying me at this meeting was a physicist friend Eric S. who was also working on the Black Knight rocket program at Alum Bay on the Isle of Wight. We had spent some of the evening talking with Adamski during which time we were frequently interrupted by impromptu callers - predominantly inquisitive mature ladies! Therefore he had kindly invited us to continue our conversation over breakfast, which was also destined to constant interruptions. As Eric and I had to be leaving to catch the ferry back to the island, true to his generous nature Adamski suggested we resume privately in his hotel room,

Of the various issues we discussed that morning, one thing impressed a disturbing memory on me. Eric and I occupied the only two chairs in the room and Adamski was perched on the side of the bed. As two professional people working on space rockets we had responded when the subject got round to future moon shots. Going by some of Adamski's lay person remarks, we were beginning to form the impression he was romancing a little now and again. But what finally left both Eric and I incredulous was a final remark Adamski made. He claimed the space brothers had told him that all our efforts to get to the moon by rocket was doomed from the start. For having reached the moon, 'the ship and its crew would never be able to lift off again'. When we queried this later, Adamski brushed it aside with some obscure reference to gravity effects. When Eric and I finally took our leave, we both felt disappointed and rather let down, for self evidently *someone* was being economical with the truth!

As with several other issues in this book, this is the first time I have publicized this incident with no wish to discredit Adamski's memory, as is evidenced by the positive technological aspects I have contributed in the Cosmic Matrix,but rather in order to keep this present report as factual as I can.

A space picture for
a charity fete

Some time after I joined Saunders Roe I had been invited by their local Horticultural Society to submit a painting for the art exhibition side show. It so happened that for some time past I had been using the air brush technique for some of my futuristic space scenes

An exhibit for visitors to a horticultural show, but as shown later in this chapter , there would
be another time and place when other kinds of visitors might cast it a quizzical look.

and had an urge to change the medium for oils, so although I was very busy with other work, a positive response to this kind request suited me. Logically, a space orientated subject would normally be unsuitable for such a meeting, even so the urge was sufficiently persuasive to override that. But a compromise was indicated. So I decided on an imaginary moonscape - bearing in mind at that time only telescopic moon photos were available, moreover it seemed natural that I should have found a good excuse to get down in pictorial form what I thought a gravitationally operated vehicle might really look like, that is according to theory.

I remember I had left the work on the picture rather late, so much so that I had begun to doubt it could be completed in time for the exhibition. But the thought of opting out while some of my workmates were working hard on their exhibits was rather daunting and to make matters worse, try as I may, I couldn't get the machine in the picture to look right. It looked too laboured and required a softer rendering. I checked the shadows perspective and the reflected light to no avail, a situation not improved by too little sleep. Later one night I stood there alone and on the point of admitting defeat, put the brushes and palette down and out of sheer exasperation mentally yelled "Help...somebody help".

The following day was Saturday and among my other tasks I had only the weekend to finish the painting. Maybe I had 'recharged the batteries' I don't know, but somehow the painting rapidly came to life, but not in a calculated way. The odd unintentional mistake made by a slip of the brush introduced an asymmetrical distortion which gave the - up until then - lifeless machine, subtle movement. The distant hills of the lunar crater took on distance, (not easily attained as there is no atmospheric softening on the moon). The true nature of the lunar surface - although speculated about - was unknown at that time, again somehow I had got it just right. Thus in a remarkably short time the painting was finished so the rest of the available time could be spent on framing.

The completed picture barely fitted in the boot of the car and I drove to work on Monday morning with a mixed sense of relief and uncertainty as to the suitability of the subject I had chosen. I need not have worried as I soon discovered when I deposited the exhibit in the lecture room vestibule.

And other visitors from where?
There can be little doubt that the size of the painting contributed to the overall attention it aroused, yet it was more than that, people, all sorts of people, even the early morning char ladies seemed to find the subject fascinating. Monday evening found the preparation for the forthcoming show well under way with entries from the staff, local gardeners and business people etc. I went along the following day and was taken aback to discover that the rather futuristic picture had not only won first prize, but had been arranged as a centre stage show piece! All at once I knew beyond a shadow of doubt, people did want to know, the space age and all it implies was here to stay. This was great and I was relieved and excited, yet how could I have possibly known what the future had in store for this hasty contribution to a local charity fete nearly two years ahead? We shall see later on.

It so happened that during this time there was an extraordinary increase of unidentified aerial activity going on in the south of England, the Isle of Wight was no exception. Among some of my new found friends were coastguards and personnel from the Plessey Radar establishment at Cowes. It was natural therefore that their equipment would be in constant use, and not only did some of the operators have remarkable tales to tell, but lay people also. Indeed I found the island a UFO fertile place with reports of such aerial activity from people wanting explanations for some of the things they had seen.

I.W. Show Catered For All Tastes

EVERY year the biggest event on the Saunders-Roe Sports and Social Club calendar is the horticultural and handicrafts show, held at Cowes.   On Saturday it was the show with something for everybody.

For men members of the club there was the fruit, flowers and vegetables section described by one of the judges as " a wonderful display of very high standard."

For women there was the handicrafts section, with a fine collection of knitwear.   The children had their own section.

One feature of the show was the art exhibition in which all pictures had been painted by Saunders-Roe employees.   Most striking were a group of space-age paintings by Mr. Leonard Cramp, and in contrast scenes of

class and runner-up was Mr. J. Rowe who carved tiny " statues " from matchsticks which had to be seen through a magnifying glass.

Entries in the horticulture classes were slightly down on last year.   One gardener grew four potatoes which weighed a total of 5½ lb.

The show was opened by Mrs. R. Stanton-Jones, wife of Saunders-Roe Chief Designer.

Trophy winners were as follows.— Saunders-Roe Horticultural Challenge

Evening News item of Nov. 2ⁿᵈ. 1959 which covered the public meeting where the author's moon painting drew a lot of attention.

This included fishermen while trawling at sea.

Perhaps not surprisingly, I was asked to give more talks and a society was eventually formed which attracted members from a local astronomical society, some with their own stories to tell. Sometimes it got a little difficult at work when staff sought me out to relay a previous night's strange story. I had more than my share of phone calls and I was invited on to the local TV programs and interviews with the press. Most of the senior staff were extraordinarily tolerant in their reaction to this. Indeed, hindsight reflection brings home to me how very little of this experience could possibly been obtained had I not moved to the Isle of Wight. Predictably I had my share of petty jealousy directed rebuffs from certain quarters. So perhaps it was inevitable that the committee of the local branch of the Royal Aeronautical Society would invite me to read a paper on the subject which was causing such a stir. Although somewhat intimidated by this I accepted, for it was considered a local honour. Accordingly a date was fixed for March 1961; little did I know that was going to be merely the first of no less than five separate, seemingly unrelated, significant events, which would occur in just as many months!

With the help of some enthusiastic team mates I built some visual aids and the lecture went ahead with an unusually high attendance. More than a little nervous I gave of my best on this very contentious subject, to a highly trained bunch of professionals from the establishment. With due respect, I anticipated that they would represent three main categories; i.e., the already convinced, the merely curious and the sceptics who had gone along for a good laugh at my expense.

After an hour I wound up the lecture allowing a little while for question time, during which we would break off for refreshment. *One and a half hours* later we were still into question time, I was getting very thirsty, given a can of beer on the stage and asked to continue. When we did eventually call it a day and assembled round the bar they were still happily arguing among themselves! They barely heard my "Goodnight", I had a wife and young family to go home to, but I shall never forget the 'well done Len' and handclasps when they realized I was leaving the hall that night! The following day I was invited into the Chief designer's office where I received his personal congratulations together with a cordial invitation to repeat the talk the following September, which of course I was happy to accept.

An alarming clandestine
meeting with strangers

One morning in August 1959, prior to the Saunders Roe Horticultural Show, the morning post brought a strange letter from a German gentleman by the name of A.Ilka, asking if I could meet him in London that day as he had something very important to discuss with me. Thinking that this had something to do with my hovercraft, I caught the next ferry to the mainland in time to catch the London train. I had arrived at Waterloo not knowing what to expect, I was in for quite a shock. As prearranged I was identified by two men, one young, over six feet tall and of rugby player/minder stature, the other shorter, perhaps in his late sixties, with white hair and a light grey suit. Both carried black briefcases and wore dark glasses. They approached me and the older man identified himself as Herr Ilka and the younger man his son. Then with furtive glances over their shoulders I was bustled away between them. I must confess that at that moment I had grave doubts of ever getting back to the island again! The situation was exacerbated by the fact that as we walked out of the station the two men conversed in German and my German was zero. Herr Ilka finally addressed me in English with difficulty and heavily accented.

However it was obvious they had transport waiting and we soon arrived at a huge black sedan. I was ushered into the rear seat between them. Ilka and the young man pulled down the rear window blinds 'to make things more private' they said and with curt instructions to the driver we were on our way. By then I really was quite convinced the next stop would be Gatwick Airport and I was seriously making plans for a rapid exodus as soon as I could! It was with no small degree of relief when we stopped at St.James Park, the blinds were wound up and I realized my fears were groundless.

Herr Ilka then became more relaxed and with gestures, broken English and his son acting as interpreter, managed to reassure me about the situation. Apart from the odd hint or two about UFOs and gravity, several hours later I realized this meeting had been little more than a 'chemistry' evaluation, which apparently checked out all right, for as we parted Ilka said he would write to me again soon. I caught the next train and ferry and was never more pleased to see the welcoming shores of the Isle of Wight!

This meeting with Ilka was to be the first and last time we would meet. He went back to East Germany and from then on our lengthy communications were by mail. He turned out to be a very courteous, brilliant, but fearful electrical engineer, who was sure he was on to a major technological breakthrough. At that time Ilka was trying to leave the Eastern Sector, so this explained the charade at Waterloo station, but it was an experience I would not wish to have repeated.

Like myself , Ilka's work involved gravity research, both theoretical and empirical and he was looking for backing in the West and a contact who could handle patents for him. He wanted me to assist in the latter and help prepare technical documents etc. But I felt that in view of our mutual interest this might have been an unwise step to take. However this became unnecessary as his rapid mastery of written English was phenomenal.

This sporadic relationship with Herr Ilka lasted until 1961, during which he spent some time in America working on his gravity ideas with private funding. When this ceased he returned to Germany disillusioned. Ilka was a fine mathematician and mildly interested in UFOs, so inevitably now and again he attempted to identify his theoretical work with the subject. I felt he was wrong in this , for his ideas wouldn't have produced a vehicle which would have been totally compatible with observed UFO phenomena, (in particular high rates of acceleration and deceleration). At one time he told me that he left East Germany with extreme difficulty and even then he had to destroy all his equipment and notes - which represented many years hard work and money - rather than let the work fall into the wrong hands.

The precognitive evidence in a
space visitor's sympathy

On an occasion Herr Ilka had said his wife was rather religiously orthodox but psychic, and she did not approve of his interest in UFOs. This is particularly interesting in view of what follows.

Although I have always refrained from quoting correspondence, I have considered it not to be imprudent in this case to include copies of the last two relevant letters that I received from Herr Ilka, so that readers can judge the implications for themselves.

The first page of the letter dated February 7th 1961 is one of several pages, the remainder discussing more general information. Particular attention should be given to the lower half of the third paragraph in which it appears the visitor conveyed a message of foreboding to Frau Ilka, although apparently Herr Ilka himself did not understand, or choose to ignore any relevance. Notice also in the fourth paragraph of this letter Herr Ilka mentioned his forthcoming stay with his son in Luton, Bedfordshire, UK and how if there was to be another such visitation it would probably be in England.

The readers attention is now drawn to the contents of the second letter from Ilka dated March 23rd in which he reluctantly declines my invitation to join our group on the Isle of Wight due to 'his not feeling too well'. Accordingly I answered the letter accepting his invitation and proposed a date we might visit him in Luton. It may have been lost in the post however, for I never received a reply. Knowing this to be his son's address I wrote again several months later asking how his father was and where I might reach him. This letter was also unanswered. Knowing how prolific and insistent a letter writer Herr Ilka had been, considering his age and sudden continued silence over the intervening years, I have concluded that something untoward had occurred to him at Luton. In which case, sad as the implication may be, within the context of the main direction of this book, one might legitimately ask the question; How did the space people know *in advance* and was Herr Ilka, five months later, enabled to keep his promise to visit me after all? In the following chapter I have tentatively laid out some details of the UFO landing at our home in which a possible link between Herr Ilka and myself is echoed.

A.Ilka
B r e m e n
Feldstr.18   7/2/61

      Mr.G.J.Cramp

      Norton,
      Yarmouth,
      Isle of Wight
      England

Dear Mr.Cramp:

Thanks for your letter of January 31th,1961.In the meantime
happened here odd things. My wife had two times contact with
persons from an another planet.

The first contact occured under miracelous circumstances. A
spirit persecuted my wife about one month long.He was audible
and feelable but not visible. In the last days of November
1960 my wife took a walk and under a bridge or subway of the
railway she heard again these ominous steps.Suddenly a man
appeared on her left side and accompanied her to the end of
the subway. He said: Du bist Martha. (Thou as Marta).My wife
said nothing but looked to his face and then to his shoes.
Twenty meters before the end of the subway the man disappeared
resolving himself into nothingress. His voice was musik in the
ears of my wife and she had a feeling of heavenly joy. From
that time the ghosts in our home ceased.

The second meeting happend January 24th,1961 at midday. This day
in the morning about 5 o clok my wife called me in the darkess
saying that a man was standing in the rear of her bed. I switched
on the light and said that she erred. In short words she meet the
man from the first meeting under the bridge which was accompanied
by a man also from an another planet. This happend on our Feldstras-
se. On a distance of twenty meters my wife recognized the man
from the first meeting.They stopped before my wife and the friend
from the first meeting greeted my wifewith the words: Iebe ohne
Ende Grenzen.(Iove without frontiers).Then he turned to his
fellowman and said: Das ist Martha von unseren Planeten.(This is
Marta from our Planet). The other man also gave my wife the right
hand in our manner and said: Iove without frontiers. My wife replied
with the same words:Iebe ohne Grenzen. Then she turned to her
friend saying: You (thou) allowed to wait me so long. Smiling he
replied: I made a great journey and was not here but this night
I was in your home. My wife said: I know it.my husband and I invit
you to visit us. As my wife spoke spoke the words " My husband"
she observed on his face sadness or grief.Then he said: Again I
make a great journey and when I come back I will visit you. My wife
replied: You will not find me for we must change the house and we
don`t know where our new home will be. The friend smiled and said:
Whereever you will be I shall find you everywhere.But let draw the
thoot otherwise you will have much pain.Then he gave his hand and
said: Iove without frontiers. The other man also gave his hand and
spoke the same words. My wife replied: Iebe ohre Grenzen.

The feelings of my wife were the same as described by mr.Adamski.
The time passed between the first and second meeting was about
two months.Therefore we expect the next meeting to be after two
months.In this time we will be in Iuton to visit our son from
March 18 th urtil April 5 th. Probable the next meeting will be
in England.

Now some words of the occurences of the meantime. Mr.B. performed
in the presence of mr.Adamski a simple experiment in order to prove
that an electrical potential repells the magnetic medium.He found

*This copy of a letter to the author from Herr Ilka concerning the bizarre
experiences of his wife was totally out of character with the usual
scientific nature of our correspondence.*

Luton, March 23, 1961
40 Linden - Rd.

Dear mr. Cramp.

I arrived at Luton in one piece and I had quiet comfortable journey. About my coming seeing you I would rather propose that I would be more pleased if you and your friends could come and visit me here in Luton. The reason for me not coming to you is that I am not feeling to well. I think it must be the english climat which does not quiet agree with me yet. Therefore I invit you and your friends to visit me here I would be very pleased to see all here for an exchange of idias. I have here my log books and different books from America which would be of much interest to you. If you can come would you be so kind and lit me know the day. I am staying here until April 4th and you and your friends are welcome any time.

Hoping to see you soon
I remain always yours sincerely

Ilka

This was to be the last communication the author had with Herr Ilka.
Further efforts to locate him were unsuccessful.

4
The Isle of Wight UFOs

1961 - 1967

During those early times UFO sightings became so prolific that one TV news commentator said "So if there are such things as flying saucers, it would appear they are watching the Isle of Wight, I wonder why?"

I would now like to place side by side with these, to me, priceless titbits, the case of Mrs.Smith and her daughter Clare Taylor of Ryde, Isle of Wight, whom I had the pleasure of interviewing and getting to know well enough to ascertain beyond doubt the validity of their claim to have seen a disc craft hovering low over trees one night in July 1961.

I took the precaution of taking along with me an independent witness, an impartial friend Alan, and my collection of UFO photographs, together with tape recorder etc. No one listening to the subsequent recorded interview could doubt the sincerity of the voices of these two women, whom I must point out were very reluctant to attract attention over the experience.

Mrs.Smith had told her husband of the sighting, but he and the painters decorating the house had laughed at her, and she resolved to keep quiet about the affair. While viewing the TV that evening she was surprised to see a Commander Mole interviewed, who together with his wife had viewed a red glowing disc through field glasses in the early hours. Puzzled, he had alerted the police. The following day the story got out and he was whisked off to the television studios at Southampton, where Mrs.Smith saw him being interviewed concerning the affair.

Thinking the Commander had seen the same object, she had ventured to telephone him to see if she could corroborate his story. Despite the good lady's protests, the Commander took it upon himself to tell the police, thus the story of her sighting came to light.

Mrs.Smith was staying overnight at her daughter's flat on the top floor of a large house which stands overlooking the whole of the seaside resort of Ryde, with splendid views over the Solent and surrounding countryside.

At about fifteen minutes after midnight, Clare decided to collect baby linen from the little roof garden above the flat, saying to her mother that they might see the lights of her husband's ship which she was expecting in from a trip abroad. Visibility was very good, they could see the mainland lights quite distinctly, yet there was cloud cover for none of the stars were visible.

Suddenly Mrs.Smith pointed out to sea where there was a row of lights apparently from a distant ship on the horizon, but this was no ship for, as Mrs.Taylor pointed out, they were *above* the horizon and rapidly getting nearer. At first, thinking it might be an aircraft, the women were not duly concerned and even when they realized it was not, felt quite certain it was some kind of hovercraft, for at that time the Saunders Roe N1 was having its trials. Anyone having seen the N1 could easily understand how it might be mistaken for a saucer.

But this thing was uncommonly silent and less than a quarter of a mile away, where it hung, almost at eye level, over some trees. Mrs.Smith said they could see the shape of the UFO distinctly, it being lit up by a row of *five* porthole lights.   She remembers this, for when thinking it might be a ship she had counted them. It is interesting to note, that if in the Adamski scout configuration there were three 120 degree sets of *four* portholes, as I showed in the Adamski/Coniston orthographic projection drawing in my book Space, Gravity and the Flying Saucer, then it is highly likely that given the chance orientation of the vehicle, an observer *would* indeed see five portholes viz. Four portholes of one set plus one porthole of an adjacent set.

The rounded dome like structure at the top was reflecting a brilliant light and the bowl shaped lower body was surrounded by a soft glow. As the object tilted now and then, they could see that it was circular, the trees below being lit up by the furnace like glow emanating from the underside.

By then the two witnesses were pretty scared and when the UFO started to come nearer, 'we backed down the stairs thinking it was coming straight at us'. But then the UFO had *tilted,* showing its glowing red base for an instant,and was gone at vertiginous speed. Said Mrs.Smith, "One moment it was there and the next instant it was going away in the direction from which it had come, like a shooting star; it went so fast it made me feel dizzy at the speed of it, and in a split second it looked just like a bright star in the sky".

Breathless with amazement they had stood there, open mouthed, for an instant before they realized something was happening at the spot the UFO had so dramatically vacated. There, slowly rising, was an *expanding ring* of smoke, or *leaves and debris;* it was luminous and hung there for several minutes after the object had gone, before it finally disappeared.

I then asked the witnesses if they could identify the craft were they to see it again, and then produced my photographs. They assured me with a laugh that they would have no difficulty doing this, for they would never forget the experience as long as they lived. Looking through the prints they said 'perhaps' yes to this and that, but when I produced George Adamski's scout ship, their reaction was spontaneous; there could be no doubt they said, they were certain this was it. It so happened that young Mrs.Taylor was a fine artist and she now, half apologetically, produced a coloured sketch she had made of the UFO. There could be no mistake about the likeness; any UFO student would identify it immediately.

We went up on to the flat roof and photographed the distant trees over which the witnesses claimed the object had hovered. These were easily located on an ordnance map, so we went out to find them. During the interview with Mrs.Smith and her daughter, Alan had said that he knew the area over which the disc hovered, as he had a friend, Mr.N. living there, and very soon we found these same trees were adjacent to Mr.N's house. Visiting him the following day I discovered the property was a large Georgian house, the ground and first floors of which were occupied by Mr.N. while two elderly ladies rented the second floor apartment.

Mr.N. seemed rather puzzled about my enquiry for it soon became evident that he did in fact have a *very* disturbed night. It transpired that just after midnight someone had not only trespassed into his garden, but for some unaccountable reason hurled a large stone through the open window, which crashed on to the floor and rolled across the room

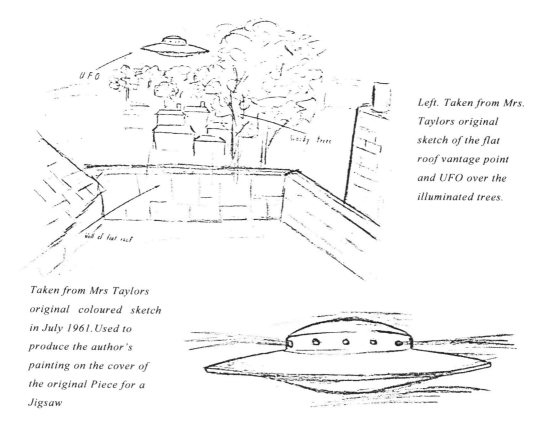

Left. Taken from Mrs. Taylors original sketch of the flat roof vantage point and UFO over the illuminated trees.

Taken from Mrs Taylors original coloured sketch in July 1961. Used to produce the author's painting on the cover of the original Piece for a Jigsaw

awakening him. Thinking he had a drunk in the garden, he had picked up the stone and gone down stairs to investigate. On going out into the garden Mr.N. was puzzled to find his dustbin lid removed and hurled some considerable distance, while the contents were undisturbed. As it was unlikely neither a fox nor a tramp would do this, bewildered he returned to bed.

A case of teleportation

Mr.N. was very accommodating and took us over the house and later the garden. The first thing I noticed was that the large stone seemed to be one of several he used as an ornamental garden edging. It measured five and a half by four and a half inches and weighed several pounds.

The sash window in Mr.N's bedroom had been open by a foot or so and outside this was girdled by a small balcony with an eighteen inch high masonry wall. Below the window was a stepped terrace, while the garden sloped fairly steeply away from the house. The height from the most suitable part of the terrace to the window was about seventeen feet. Now I know of no athletic cricketer who could have - in the darkness - had the strength and skill to hurl that stone with the required force and trajectory, up and over the balcony wall and neatly into the open window. I was sure it was impossible and said as much. Of course Mr.N. had already reluctantly arrived at the same conclusion, with the priviso, that the person intended to smash the window. But this still didn't explain *how,* for it would have required a world shot putter par excelence to have achieved it.

And there the matter would have ended but for the fact that a few weeks later Alan happened to mention that he knew the two ladies who occupied the floor *above* Mr.N. and visiting there he had out of interest asked them if anything unusual had occurred that week. They looked at each other rather shaken, and somewhat diffidently confided in him because they 'didn't want to offend Mr.N.' Apparently that very same night and at the same time, *two* similar stones were dropped into *their* bedroom with a thunderous crash and frightened the life out of them.

Now it seems we had *some* athlete on our hands, for the height from the terrace to the ladies bedroom window was no less than twenty seven feet. But that became rather academic when I learned that the two dear souls *never* slept with the windows open and their bedroom door was *always* locked! Later Mr.N. replaced the boundary stones in their located places, leaving one bewildered Ufologist to ponder, not so much on whether this thing occurred, nor even how, but why?

Now this case was extremely interesting to me, for now I was not only in possession of the Adamski photograph of the scout craft (which together with the Coniston photograph I had been able to produce my orthographic scale flat elevation drawing), but I now had a very good colour representation of the same ship portrayed by a first class artist. Moreover Mrs Taylor had told me she thought the angle presented by the craft she saw was nearer to that of the Coniston photograph, and going by her rough sketch, it was even less angle than that. Therefore it was a simple matter for me to orthographically project from my flat sided master elevation drawing to a *new* perspective drawing having an angle finally agreed by Mrs.Taylor. Then by completing the details from Adamski photographs, together with the new colour guide lines, I was able to provide a painting. Thus for the very first time , we now had an authentic *coloured* picture of an Adamski type scout ship, that is, with the exception of the spherical landing gear which Mrs.Taylor did not notice at the time. Therefore I omitted these in the painting, which was later used on the dust jacket of my book, Piece for a Jigsaw. In those days of course we didn't have computerized graphics, but although the orthographic process may have taken a good deal longer, it was just as accurate nonetheless.

This sighting in July 1961 heralded the beginning of the extraordinary events which would overtake me during the following years and a collective review of these facts now, reveals self evidently, UFO activity was coming closer to me. It should be borne in mind that both before, during and after Mrs.Taylor's sighting I was very busy putting together the manuscript for my book Piece for a Jigsaw, when during the following month of August my life time study of UFOs seemingly brought one to my door.

Now it begins

Late one pleasant summer afternoon shortly after my interview with Mrs.Smith and her daughter, I was standing at the bar of a local inn in Yarmouth, I.O.W., enjoying a drink with a friend Jack. We were there to make final arrangements with the proprietor for the use of a large room for our society's AGM during that evening.

At that time the lounge was comparatively empty, with only Jack, myself and the proprietor at the bar. Having already spoken to him and joined him with a drink, it was inevitable the subject of UFO' would ensue.

Jack was standing on my right side, and noticing him glancing to my left, I looked around to find that we had been joined by another person. He was of medium height, perhaps five feet ten inches and looked in his middle thirties. As near as I can remember he was dressed in a beige coloured suit or lightweight summer jacket. He had dark hair and was lightly tanned, though this is difficult to describe, perhaps something more like a light coffee. When the stranger smiled his eyes sparkled, his perfect teeth, in contrast to his facial colour, were very white. Unfortunately by nature I am inclined to be a little reserved when unexpectedly approached by strangers. However I was immediately reassured by this person's manner when he joined us in our conversation.

He politely said he could not help overhearing about our UFO society and how interested he was in the subject. He said this was his first visit to the Isle of Wight and how he thought we were so fortunate to live in such a lovely unspoiled place. He really would like to come again sometime etc. Eventually the stranger bade us farewell and we resumed finalizing the meeting.

About a week later I received a short extraordinary letter from him in which he reiterated how pleased he was to have made our acquaintance and again expressed his love for "your beautiful island". Then totally out of context he said "I wonder what you would have said had I told you I come from the planet Venus?"

Granted there are some strange people about, but I really think there was something different about this individual. The experience was a pleasant one and regretfully cannot be effectively conveyed.   As also is the following incident which has remained   as paramount as it was mysterious, for the rest of my life.

The CE3 landing at
Gardeners Cottage

On August 12th 1961 I had received a newsletter from New Zealand which had somehow got laid aside. Having arrived home fairly late on the evening of 19th August, I came across this letter with some others, but due to the late hour felt too tired to read them. However I still had a nagging persistency about the newsletter from New Zealand and I decided to at least glance through it. The contents of the letter proved to be intriguing, referring to a world wide experiment in telepathy due to be carried out the *following day* at twelve noon!

Now coincidentally at that time my younger brother John and his family had been having their annual reunion holiday with us, but due to building work underway in our house, they had been sleeping in a large marquee on our rear meadow, so that by the time I read the newsletter John and his young family had already retired for the night. As for me I also went to bed and slept the whole night through quite unaware of the extraordinary happenings which would then occupy the rest of my family, as told to me the following morning. Apparently as John walked towards the tent that night - some twenty yards from the house - he thought he saw flashes of light towards the end of the meadow. Thinking the lights were shining through the trees from a distant house he had barely registered it. Then, realizing that the lights were in the wrong direction for the houses, he had looked more intently but saw nothing. He then dismissed the matter and retired to bed.

The following morning found all sorts of members of the family interrupting each other with their own story to tell, therefore of necessity I must relate the events in the order in which they occurred, rather than in the order they were related to me. But first a little geography is required.

Our twin daughters Jane and Sue occupied a bedroom immediately above their brother Gary's room on the ground floor of the house. Their respective ages at the time being, the girls six years and Gary nineteen. Also it should be pointed out that the windows of their rooms were situated one above the other. Jane's bed upstairs with the foot beneath the window and Gary's likewise immediately below.

The site of the property was essentially rural with some six and a half acres of surrounding garden and meadowland. On the same parcel of land my family and I had

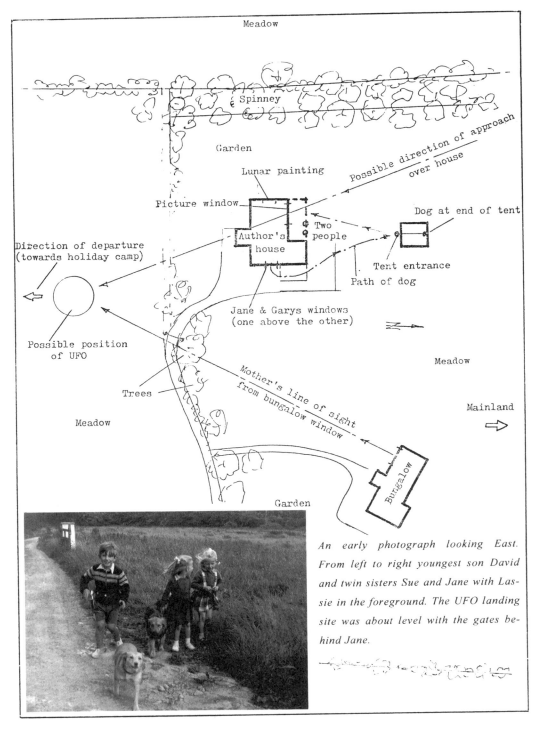

Meadow

Spinney

Garden

Lunar painting

Possible direction of approach
over house

Picture window

Dog at end of tent

Author's
house

Two
people

Direction of departure
(towards holiday camp)

Tent entrance

Path of dog

Possible position
of UFO

Jane & Garys windows
(one above the other)

Meadow

Trees

Mother's line of sight
from bungalow window

Mainland

Meadow

Bungalow

Garden

An early photograph looking East.
From left to right youngest son David
and twin sisters Sue and Jane with Lassie in the foreground. The UFO landing
site was about level with the gates behind Jane.

Site plan of the shared UFO incident, precipitated by the family dog, which independently
involved no less than nine people at the author's home, which took place on the night of
August 20th 1961. This plan can be viewed in conjunction with the photos overleaf.

View of author's cottage looking west. The post in the right-hand foreground marks the entrance to the bungalow. The landed UFO was positioned level with the gate and to the left of the photo. Daughter's bedroom window can be seen at top right of building with Gary's window below. The UFO approach would have taken it over the trees and house down to front meadow, departure being to the extreme left.

View looking along the estimated UFO flight path as shown on the site plan. This was coincidentally photographed without knowledge of the UFO incident. Brother's tent was positioned to the extreme left of the picture and the two figures were standing at the nearside of the loggia's central pier while looking through the picture window.

built a bungalow for my parents and sister some one hundred and fifty yards from our house.

When the twins were only a few months old we had acquired a lovely little mongrel dog called Lassie who became devoted to them and as they grew up they would encourage her to sleep on the foot of their bed. That was the situation on the night of August 19th when the girls retired with Lassie on the foot of Jane's bed. A little after 2 am Jane was awakened by Lassie excitedly whimpering and yelping and leaping up at the closed window as if trying to get out.

Then Lassie jumped off the bed and made for the door followed by Jane, who opened it to let the dog out, whereupon Lassie scrambled downstairs, leaving a somewhat puzzled and not too pleased Jane, who promptly jumped back into bed blissfully unaware of the nocturnal drama she had just unleashed! Lassie had then awakened an equally startled Gary by lunging at his partly open bedroom door and on to the foot of his bed in order to get to his ground floor window. Thinking a marauding fox was the cause - as was often the case - he had let the dog out of the window and like Jane got back into bed.

My brother and his family were then awakened by Lassie who was whining and in a worried state trying to get into their tent. It should be stressed that this was most unusual as Lassie was normally quite fearless where foxes or the like were concerned.

Having made access into the tent, the dog went to the far end away from the house, jumped on to a bed containing one of the children and in the torch light continued to whine and stare in the direction of the house. John managed to quieten the dog and noted the time, it was just after 2 am. He then thought he must have dozed off, during which time he had a very vivid dream. In this Lassie once again became frantic and carried on as before, again looking terrified in the direction of the house. This time my brother decided to investigate the cause of the disturbance, got out of bed and went to the tent flap facing the house.

John said the house was in complete darkness save for the large picture window about eight feet by five feet which looked out through a long loggia. The window appeared to be strangely illuminated and near it stood two figures. They were of normal height, one taller than the other, who had one hand on the other's shoulder. They were dressed in one piece suits of a shimmering smoky blue material, tight at the neck, wrists and ankles. The two were obviously looking in to the picture window and John found himself thinking "who the blazes are they and what are they looking at?" Almost as though in answer to his thought, my painting of the lunar landscape hanging on the lounge wall appeared to light up. John thought they were interested in this and were looking at it in an appraising manner. There the strange dream-like experience faded until he remembered it in the morning.

Literally while I was listening to all this my elder brother Alf, who lived at another part of the island arrived, said "Hello" and not wishing to interrupt what was obviously an interesting discussion, stood by listening, no doubt thinking that *his* news had already been anticipated. However as the discussion ensued he began to suspect that he knew something that we did not, so looking at me he said, "Len, have you seen Mum this morning?" then responding to the hesitant look on my face he said "Well I think you should".

As stated, my brother Alf, my two sons and I had built my parents house ourselves and were quite proud of it, although the job had taken over three years to complete with many problems along the way. Now, I had cause to remember that time when just before dusk I had stood with hammer and pegs in hand ready to set out the site for the bungalow in my Mother's presence. She had been staying with us and would have to be leaving shortly.

She asked if I could "turn it round a little", so that when she and my father got up in the morning they could look over to our house "which would be nice". I grinned and said "that's OK Mum" while acutely aware of several difficulties that might bring. Little was I to know how important the fulfillment of my mother's wish would eventually be.

Now, as I stood looking at her that morning, my mind already keyed up by the night's happenings, she related her side of this remarkable story. It should be stated, at that time my mother was seventy four years old, of sound mind and certainly knew nothing about space ships or flying saucers.

At 2.12 am being unable to sleep, she had noticed her bedroom seemed to be unusually bright. So as was customary on such occasions, she got out of bed to look over towards our house. There, just above ground level behind the trees which stood in front of the house, she saw 'a most beautiful golden object like a huge, lit up, bell'. It was moving slowly upwards. It appeared to rock a little from side to side until it was above the trees. Next there appeared above and around it a 'brilliant silvery ring of light' then the object had 'appeared to fall' at a terrific speed (towards the south) and my mother held her breath 'thinking it was going to crash'.

Now although my mother had sometimes shown signs of being very psychic, it was a faculty she played down or even denied. Neither was she overly religious, yet that morning she looked at me with misty eyes and said "I don't know what it was son, but it was so beautiful that I found myself thinking, if ever Jesus came again that's the way I think he would come". That was a most extraordinary thing for an old lady to say. I felt terribly inadequate, gave her a cuddle as I thanked her and returned to the house with no small lump in my throat! I shall always remember and be thankful for those words which I like to feel were a very natural, unbiased, subconscious classification. Eventually when shown the Adamski scout photo she said "Brightly lit up it would be the same".

Later that same morning while the whole family were still debating all this, I received a phone call from a friend to tell me of a UFO sighting witnessed by a guest staying at his mother's boarding house at Ryde. She too, being unable to sleep that same night, had seen a very large orange 'shooting star' in the west, i.e. in the direction of our home at Yarmouth, the time was about 2.15 am. My brothers and I scoured the land around our cottage for any signs of ground effects and in the area indicated by my mother we found tell tale signs of whirligig patterns, but due to the shortness of the grass we decided that this was inconclusive.

My brother John did take part in the experiment set for the following day 12 noon, 20th, but promptly fell asleep and had a strange dream. In this he was looking down onto our home from a fair height, then the scene changed and the entire terrain became a vast waving sea of brilliant red poppies. As he watched, some of the poppies began to line up and form regimented patterns, and that was all he could remember. But after he and his family returned to their home in Brentwood (on the 29th), I received a letter from my nephew Johnny, then aged nine, dated August 31st 1961, saying:-

Dear Uncle Len,

I saw something in the sky on Tuesday night at 10.30 pm. I was getting into bed when I suddenly heard my friend in the garden next door say, "Look Mum there is something in the sky" I looked out of the window, all round the sky, but could not see anything, then I saw something. It was a round object, its colour was light red. It stopped for a while, then it went along quite fast, it glittered on top and round the side. I went downstairs and told Dad and we both went outside but it had gone. Dad said it could have been a flying saucer and I am very pleased to have seen one. I have drawn a picture of it for you in colour, but I could not do the glitter.

Same view of the cottage looking west as it is now in a cul-de-sac. All that can be seen is the end gable beneath which is the window of the twins bedroom.

Early the following September, a local fisherman was interviewed by a journalist friend of the author's concerning a bizarre phenomenon he had encountered while drifting off the Isle of Wight Needles (which is only a few miles from our home) *on that same night of August 20th*. His boat was equipped with radar and he detected a moving target on course with his position. Without warning his boat lurched forward gaining speed. Despite the fact that the evening was calm and the sea state slight, the sea around him rapidly built up to three foot waves, local, too local. Coincident with this, the radar target stopped and his boat began to move in ever decreasing circles and the compass was spinning 'like a top' and he was very alarmed.

After a few moments of this mini nightmare during which he could do nothing, the motion stopped, the sea began to settle and everything returned to normal. Did he remember the time? Yes, somewhere between 2 and 2.30 am!

A belated important sequel

Recently at a family reunion, we had been reminiscing about this extraordinary affair when our daughter Jane, now a mature lady of forty two, broke in, pointing out an important omission concerning that night of 20th August 1961, about which, remarkably enough, I had been unaware all these years. I knew of course it had been she who let Lassie out of the bedroom, but what I didn't know was that as a six year old little girl, she must have tried desperately to tell me the full story. But sadly somehow, in the ensuing excitement, she must have been brushed aside. Over the years she assumed I already knew. So it was with my apologies and gratitude,she was now at last able to tell her *full* story.

Jane said she had indeed been woken by Lassie whimpering and yelping and scraping at the window, but this had been accompanied by a 'humming noise' seemingly coming closer and above the house. As it got louder so the frantic dog responded. Jane could still remember she thought the noise was caused by 'some kind of helicopter, but not nearly so loud'. But whatever it was she felt it was very near as she sensed 'a kind of throbbing vibration all over the room'. Then Lassie had leaped off the bed and made for the door followed by Jane who let her out.

As Jane told me this, I couldn't help imagining the frustration she must have quietly endured when, as a little child, she had desperately tried to make an over excited parent listen to her full account of an episode in which, after all, *she* had helped precipitate and to take a major part.

Another unexpected
supportive sequel

The following additional incident occurred so long ago that I now feel free to disclose it and do so for the first time due to its supportive relevance. In February 1967 the Isle of Wight UFO Society (founded in 1959) of which I was a founder member, held an exhibition in the town of Newport. There were many interesting exhibits, including an astronomical section which incorporated some very useful models of Dish Aerials, kindly loaned to us by Plessey Radar Ltd., of Cowes, which is about ten miles west of our home. This exhibition was very successful, attracted a lot of attention and was reviewed and complimented by the media, which included national dailies and television. During the evening a young man singled me out because some years ago he had become interested in UFOs. He was a top grade Radar operator and his job very often entailed him working at night. As we were both technicians we immediately struck up a rapport and he confided in me the following story. He said that some of the equipment they were using was so sensitive "you could pick out the movements of London busses with it" That was interesting, but what he said next sent my pulse rate up a bit. Apparently some time in the summer a few years back, he had been on night shift, when he suddenly spotted something on the scope which demanded his immediate attention. He had a target which moved *very* fast and erratically, then suddenly stopped! Neither birds nor aircraft usually do that, so by this time he was very puzzled. Of course he was aware of UFO reports, but like so many others, he was apt to have little sympathy in that direction.

The target had repeated that manoeuvre several times, then suddenly was joined by another, whereupon they 'went off the screen' in a *westerly* direction. He computed the speed and found the objects were going at over three thousand miles per hour! The young man confided that some of his colleagues had seen similar targets on their screens, but it had been regarded as 'sensitive' information.

He kept to his promise and looked up his notes about his 'experience' and saw no harm in telling me the date. It was August 20th 1961....at 1.55 am!

As is shown elsewhere in this book there were other events - both before and after those presented here - but I feel the occasion at our home on that August night is important in as much as it, in common with a whole range of other similar world wide incidents, included a fairly comprehensive spectrum of physical effects, including audible and visual, but in this instance not only shared by quite a few people, but also by our beloved pet dog Lassie!   An inclusion which rather negates the 'shared hallucination' theory desperately held on to by some people. Neither can we sensibly ascribe this event to a hoax, for if indeed it were to be such, then it must surely have been one of the most elaborate and pointless hoaxes of all time! Neither would it be proper for me to offer my own tentative explanation for this event here, suffice to add, for the discerning reader there is a clue.

A memorable visit to Wales
The forestated repeat lecture at the Saunders Roe branch of the Royal Aeronautical Society in September 1961 proved to be almost a carbon copy of the first, except I now had time to polish up the format and had new data to present. Perhaps predictably the attendance was even greater, as some people were there who had missed the first program. However this occasion turned out to be rather different, for not only did the committee treat me to a splendid meal, but presented me with the news that we had been voted 'best lecture of the year' by the local branch committee and some of their opposite numbers from the mainland who had come to hear it for themselves.

Receiving this commendation required giving the lecture again at a venue chosen by the Central Royal Aeronautical Society committee and this was to be in Wales to a joint meeting of the RAE at Bristol and nearby BAC Royal Aeronautical Society branches. Being a joint gathering required the use of a large canteen hall at the BAC, so we required pretty good sound and electrical back up. This together with my own demonstration equipment was rather formidable, and I was glad to accept a kind offer from a friend Peter Wrigglesworth to use his small van for transport, which at the time happened to be in a garage awaiting an engine change.

The lecture was scheduled for December 10th 1961 and I was getting anxious as Peter's van was still unavailable. Eventually this was sorted out and we made our way to Wales comfortably at a respectable speed. On arrival we were unexpectedly ushered into the BAC boardroom to meet with the Chairman and Wing Commander from the RAE together with Dr.S, a well known physicist who would be taking the chair.   I was rather apprehensive about this, for particularly in those days I could hardly expect him to be a UFO sympathizer.

However the lecture went ahead as scheduled and Peter and I were relieved by the very reassuring response. Dr.S's summing up couldn't have been more thoughtfully supportive if he had been an ardent Ufologist for years! Among other things he said "some of the more obtuse aspects of classical physics had been admirably demonstrated", and I was very grateful I had been able to build a model or two to help the evening along. Our hosts were very enthusiastic and generous and treated us to the customary evening meal at our nearby hotel.

Coincidence or sabotage?
Out of respect for the newly installed engine in Peter's van, the following morning we made a very early start to allow sufficient time to get our ferry back to the island. All was going smoothly as we drove through the town of Bath, when suddenly there was an horrendous noise from the engine bay. We came to a stop, alighted and unbeleivingly

gazed back along the road at a thick emulsified mixture of oil, water and co-mingled smashed piston rings, con rod and assorted bits of crank case. The new engine had blown apart! We had traveled too far to return to Wales so we had little alternative but to secure the van and its contents and make other quick arrangements to get back home.

A week later Peter retrieved his van from the mainland and had it taken to the garage who had installed the new engine. He was present when they took the ruined engine out. They lowered the sorry sight on to the work bench and started to dismantle it and were amazed to find that every tab lock washer from the piston big ends was missing. The nuts had been put on little more than 'finger tight', causing a big end to become detached and the piston had gone completely through the crankcase wall The mechanics were dumbfounded, they were at a loss to see how such a thing could happen, they had personally taken the new engine from its wooden crate to install it in the van. Subsequent inquiries with the manufacturers failed to throw any light on the mystery, they said due to assembly line procedures, such a thing was impossible. But there it was, it had occurred.

Eventually they gave Peter another engine and while reminiscing one evening about this episode, Peter looked thoughtfully at me and asked "Without lock washers, how long into the trip to Wales would you have given the big end anchor nuts to unwind?" The thought had already occurred to me several times, so I said "Logically I would have guessed less than an hour or so", "Hmm" said Peter "in which case we were lucky to get there at all!" We looked silently at each other for a moment, then Peter went on "Bearing in

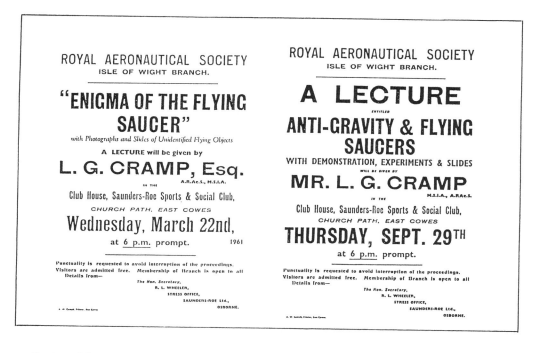

Copies of the posters billed at Saunders Roe (later the British Hovercraft Corp.). This lecture in March 1961 and its requested repeater in September 1961 earned the Royal Aeronautical Society UK 'The Best Lecture of the Year' award, resulting in further presentations to various branches of the scientific community including a joint gathering at the RAE and BEA establishments in Wales in 1962. The extraordinary events which accompanied the Wales lecture and afterwards, have never been satisfactorily explained.

mind the Wales branch committee had already reshuffled their existing annual lecture program, yours would have almost certainly been cancelled!"

As we shall see, no less than twenty two years later I would once again be involved with an identical *engine breakup* scenario. That also would be at a *public gathering* involving another of my endeavours, neither would it be on the Isle of Wight, but none other than *Bristol* airport. How far is that from **Bath**? Less than three miles! As we have seen Carl Jung called this kind of repetitive phenomenon *Synchronicity.* I trust the underlying theme in this present book and The Cosmic Matrix will go some way towards explaining it a little more satisfactorily than that.

The vanishing UFO and the
malfunctioning camera

After the event which occurred at the author's home in 1961, there were other incidents which I investigated. The following is one of them.

It was 3 am one cold December morning in 1964 when Mrs.Joan Pyner (a relative of the author's who does *not* believe in flying saucers) was awakened by "a noise like a helicopter coming over the house. She said the noise got" slower and lower in pitch" as the source came closer towards the meadow which adjoined her garden. Mrs.Pyner likened it to "the rotating blades of a helicopter coming down, though I was puzzled by it not being as loud as I thought a helicopter would have been, that close, and it *was* close, I knew that". Thinking whatever it was might be in trouble, she got out of bed and went to the window to investigate.

The grass in the meadow was quite visible, and there, only some twenty five yards from the house, sat the 'thing'. It was black, "I couldn't see the exact shape for certain, but it would be the size of a large armchair", she said.

When I asked if in fact it may have been a cow, Mrs.Pyner laughed outright, "Don't you think I can tell a cow even in *that* light?" she asked.

In order to obtain a better look, she had thrown up the sash window which made a loud squeaking noise. Whereupon the dark mass simply vanished. The witness had no idea if in fact it moved at all. There was no noise, nothing, it simply ceased to be there! It is interesting to note that Joan's home is at *Whippingham* and less than a quarter of a mile from where Mr.Jarvis saw the Adamski type cigar vehicle in 1954.

The following morning Joan phoned me at B.H.C. (the British Hovercraft Corporation) to tell me of the extraordinary event of the night before, and as my journey home later that day took me very close to her home, I made arrangements with a friend Mike A. to call there. Mike was an expert photographer and due to the failing light, thought it prudent to bring with him a camera with automatic flash.

Arriving at Joan's home I suggested that it might help if she could position herself at her upstairs bedroom window, while Mike and I stood in the adjacent meadow. The wall of the house in which the window was situated was only about three or four yards from the boundary fence, and after being directed by Joan to the approximate spot where she had seen the object resting, we began to search the area with our torches. Then after a few minutes Mike spotted an indentation in the soft grass. It was roughly rectangular in shape and measured approximately eight inches by five inches. The sides were fairly clean cut and the impression was about three inches deep. Some minutes after that we located three more, forming a square with sides of approximately six feet.

Highly intrigued by this we hastened to get Mike's camera ready. Meanwhile Joan was kindly preparing some hot tea as the early evening air was very chilly. With the short notice we had , I was unable to borrow the firm's Geiger Counter , but I was fairly content about the prospects of some useful photographs. However this was not to be, for although Mike had never experienced trouble with the camera flash unit before (in fact he had only used it several days earlier), we were unable to get a single successful shot. Moreover, despite the fact that there were no cattle in the meadow that night, the following day there were, so the indentations were completely obliterated.

Subsequently Mike was able to check the flash unit and it worked normally, also as there had been at least some light, he eventually developed the film only to find that those particular frames were blank. Strange as it may appear now, the possible relevance of this malfunction involving a 'close to home' UFO event , as with some of the others in this book was not fully realized by me until much later. At the time I had merely put this 'hiccup' down to coincidental failure of the camera equipment. However, in the light of many more recent similar cases, I think it fair to emphasize this, lest it be suspected that I tend to over interpret the 'normal' in terms of the 'abnormal'.

A title for a new book
and guests invited

Towards the middle of 1965 found us completing my new book, which due to the fact that it was to be published privately, had entailed a considerable amount of work. This included not only the text and formatting, but original artwork for many sketches and diagrams, test rigs, photography, dust jacket design and art work etc. It was some marathon, but as always, with the help of my wife and several friends, we eventually made it. For some time we had been trying to decide on a suitable title for the book, but without success. Indeed I often found myself wishing I had the benefit of a publisher's experienced opinion. Then one night I had one of those by now familiar vivid dreams in which I walked into an office and there confronting me was an individual sitting at a desk. He was busily engrossed with something on the top and as I approached, it became apparent that the entire desk surface was covered by a partially completed large jig saw puzzle. With one piece still in his hand, the man looked up at me and smiled, it was Waveney Girvan!

The following day I recalled the dream and suddenly realized I had the answer for my book. This was the title *Piece for a Jig Saw* received and understood with all due thanks to its proposer. At the time of the dream, by then long out of touch, I had no idea that Waveney Girvan had in fact died in 1964!

One of the people who kindly assisted me to meet the publishing date was a young lady named Yvonne R, who was also employed at B.H.C. At the beginning of the work she knew very little about UFOs, but as she became acquainted with the material she was reading and pasting up, she became more intrigued.

Then on one occasion she experienced a paranormal phenomenon in which the whole of her bedroom became lit with a beautiful ethereal light accompanied by the strains of pleasant music. I had known for some time that she was very orthodox religeously inclined and wasn't unduly surprised about this.

Being a rather excitable, jovial type, on one occasion Yvonne had exuberantly exclaimed "all right if these space visitors exist they are welcome to pay me a visit anytime". As with the incidents which accompanied the affair at my home at Gardeners Cottage, it is important to emphasize that the events which followed Yvonne's innocent enough remark were only pieced together by the author several months later. In view of what follows, some ardent UFO researchers might find my attitude extraordinary, but I can only defend my lack of immediately identifying a possible link to the following situation as a measure of my forestated guarded tendency to explain the unknown in terms of the

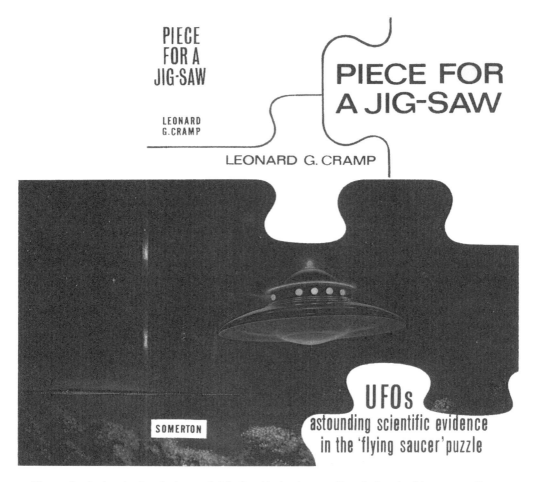

PIECE
FOR A
JIG-SAW

LEONARD
G.CRAMP

PIECE FOR A JIG-SAW

LEONARD G. CRAMP

SOMERTON

UFOs
astounding scientific evidence
in the 'flying saucer' puzzle

The author's dust jacket design and title for this book were directly inspired by extraordinary personal experiences. But why in 1966 was production of it ruinously sabotaged so that as with 'Space, Gravity and the Flying Saucer' before it, the entire first edition had to be pulped. And why in 1979 did visiting Russian scientists order half a dozen copies, then send for a dozen more the very next day?

known. Even when this sometimes borders on the head in sand syndrome!

Yvonne's home was a semi detached house situated on the main road of a small village. She had lived there all her life with her parents Mr. And Mrs.R, who were retired. Arriving at the house one day to deliver more paste ups, I was invited in for the usual cup of tea, during which they told me of a domestic incident which had occurred during the night several days previously.

At approximately 3 am. their neighbour Mrs.Sullivan, a lady in her middle forties, was disturbed by a knocking on her front door, which at the semi detached property shared the same porch as the front door of Yvonne's house. The awakened neighbour, not having a telephone, hurried to the front door half expecting to find the police delivering an urgent message concerning her husband, who was at that time ill in hospital. Before opening the door she had naturally enquired who it was at that time of night. The poor woman was both relieved and alarmed when a voice said "We just want to talk to you". Realizing it wasn't the police and maybe a prowling drunk, she told the visitor in no uncertain terms to clear off or she would get the police. Mrs.Sullivan had then gone into the front room and peeped through the curtains, but despite the street lighting there was nobody to be seen. Yvonne's neighbour had related this to her the following day.

At that time I knew that being retired, Mr.R. liked to tinker around with all kinds of gadgets now and then. So when I called at their home several days later I was not surprised to find a sense of good humoured mystery being shared by the three of them. I was ushered into the kitchen and with a "let him try it for himself" remark from Mr.R, I was told to open the back door. Going along with the situation I grasped the brass door knob, turned it and all hell broke loose, there were bells ringing everywhere.

Having got over the shock, I was then informed that since my last visit, someone had been knocking on *their* front door at three o'clock in the morning, and on the last occasion when Mr.R. went down quickly to investigate, he heard someone in their backyard trying to turn the back door knob! Hence his DIY burglar alarm system. Perhaps not surprisingly there never was the opportunity to test it out, other than on me.

Mr.R. reported the incident to the Police, but short of an overnight vigil there was nothing they could do about it. Then things began to take on a more sombre aspect. Mr.and Mrs.R. occupied an upstairs bedroom at the front of the house and one night, at about the same time as the other disturbances, they woke to find a spot of light moving around their room. Mrs.R. was awake first and sensing that the light was directed through the front window, got out of bed to investigate. Immediately opposite their house was a roadside garage with petrol pumps and to the right of these was a telephone kiosk and a small bench seat, all being well illuminated by the street lights. There were two figures in the open kiosk seemingly making a telephone call, several more stood around outside, and three more sitting on the bench looking up towards the window through which one of them was flashing a bright torch beam onto Mrs.R. as she stood there. Annoyed by this she pulled the curtains and returned to bed telling her husband that the source of the trouble was a group of youngsters "probably Army cadets on manoeuvres". Later when I asked Mrs.R. why she used that particular description, she said because they looked like small boys, dressed in uniform complete with shoulder packs and wearing helmets!

The next time I visited the R.s things had taken a more mischievous turn with Mr. and Mrs.R. being pestered with poltergeist type knockings in their room and in particular on the wall at the head of their bed, the wall being an outside one, and inaccessible from the small gap between it and the next pair of houses. Until then I had tended to accept these goings on in the same light as did the R.s, that is drunks and/or mischievous lads. Granted it was all coincidental close timing, but poltergeist type phenomena? I had been acquainted with the same pattern before!

Eventually things returned to normal until about a year later Mr.R. sadly died from cancer. Of course I never revealed my eventual suspicions, which with the benefit of hindsight now looks more evidentially conclusive and the case is best left at that. Suffice it to say the reader would do well to bear this case in mind as we now move on to probably the most bizarre instance of them all.

A repeat performance

In the summer of 1966 we eventually published my book Piece for a Jigsaw which was essentially an update on Space,Gravity and the Flying Saucer, but now with even more evidence of a technical kind which I thought more people might like to be aware of. I thought it fair to say it offered evidence which is difficult to refute. The only alternative would be to proffer a fantastic coincidence of such astronomical odds as to be even more untenable.

The book was eventually sponsored and published with the help of a very dear friend. This step being taken due in part to increasing awareness of unofficial resistance towards meaningful UFO research programs and the suspicious events which overtook my first book after its publication. In this way we thought we would be able to keep a closer rein on the situation. Our preparations were thorough. We chose one of the largest and best printers in southern England and were prepared to accept the incurred added expense believing there was little that could go wrong. But it did in the most bizarre way! I suppose I should have known it from the very beginning, when with great excitement I had received the first few packs of twenty five copies, only to find that through careless packaging they were ruined. At first I had to accept it philosophically, but each pack had been wrapped with sheets of brown paper and tied with coarse *binder* string with many knots. In much the same manner as one would receive a bundle of newspapers! Thus the hard back covers of every outside book was severely dented by the knots and string. I would complain about this and be glad to receive other packages more carefully wrapped, the damaged ones we had to discard. But that hope rapidly began to fade as I thumbed through the first copies. Despite the fact that my wife and I had checked and rechecked the galley proofs again and again, I found error after error. Not just the odd word or two, but *whole paragraphs* out of order. There were technical misquotes that were bound to - and did- bring sneers and denials from some aloof academics in the establishment game for an easy bite. One elementary instance will suffice, viz. Acceleration happens to equal 'mass times *change* of velocity'. But in the book the term *change* was conveniently omitted, giving'acceleration equals mass times velocity', which of course gives units of *momentum*. Obscure to lay persons, but very convenient for a doubting scientific fraternity to whom I was largely addressing the arguments. To an engineer such a *coincidence* is telling, it could be described as either clever or just plain damned silly!

Many pages had been printed with mixed old and new type fonts, so the print on opposing pages looked completely different, with one page crisp and black and the other faint and smudged.   The corners of bunches of pages were folded and then guillotined with the resulting christmas tree effect when unfolded. Whole bunches of pages were crimped and folded down through the middle. The expensive four colour dust jackets showing my laboriously produced rendering of Mrs.Taylor's Adamski scout sighting had been damp packed without protective sheets in between (obligatory in those days) causing most books to be literally glued together, it was a veritable disaster and we could hardly believe our eyes. Sometimes in sheer exasperation with my own inability to factually describe this abortive product, I would quote the situation as being tantamount to that of a well known, expensive, but excellent builder putting up a house without mortar to bond the bricks!

But of more importance was the fact that the competent respectable image I had tried hard to create was entirely lost. To the lay public we had just another cranky Flying Saucer bookon our hands......six thousand of them! Of course we tried in vain to rescue a few copies, but in the end they had to be pulped. Extraordinarily enough the directors of the company responsible for this charade seemed to be as equally dismayed and made tentative excuses for how it might have happened. They offered and did reprint the whole consignment. But in those days a book took months to print, so by that time we had lost the advantage of our company's expensive limited advertising. True we now had a nice book, it was extremely good value at its price, and wherever it *was* shown it sold immediately. Despite this, strangely the big book sellers didn't want to know and a member of my family who held a responsible position in the industry hinted that 'pressure' may have been brought to bear. At the time of writing when spy catcher books are being banned, I would not be at all surprised !

Eventually without the sufficient means for distribution the remaining copies were ready to be pulped when they were sold off cheaply to a friend, who sells them still to this day. This is an absolutely true account. For the sake of brevity many of the supporting details have been omitted, but all the relevant documents are safe and under wraps. Thus we have two books out of two ruined. Not bad going in terms of...coincidence is it? But of no less importance is the question; given the likelihood of such an implied scenario in this and the charade at Wales, on whose authority was it carried out, and by no means least.....how?

Contactee with a difference

Sometime after the Isle of Wight UFO Exhibition my colleagues and I received an invitation from a Mr.Blake to meet a person I shall call Mr.Jones, who alleged he had a UFO close encounter experience. Intrigued we subsequently met them both. Mr.Jones appeared to be in his early forties, of medium height and with sandy coloured hair. If I had been asked to describe the most outstanding thing about his appearance I would have to say, his quiet patient manner and his very pale, penetrating but kindly eyes.

Having settled down in the lounge of one of our party, Mr.Blake, a tall older man, volunteered to address us first on behalf of Mr.Jones. He explained that this was an agreed arrangement between them in order to "conserve Mr.Jones energy, as he was inclined to tire easily".

Mr. Blake gave us a brief résumé of how they first met. It appeared that some years earlier he had been a professional therapeutic hypnotist and at a public meeting Mr.Jones had been selected from the audience to take part in a demonstration. It soon became evident to Mr.Blake that Mr. Jones had very special powers and he became additionally interested when the man said he had seen a flying saucer.

Accordingly Blake arranged for the two to meet and Mr.Jones revealed more about the encounter. At that time he lived on Hayling Island and was in the habit of taking early morning walks on the beach. Of the sighting, he remembered unexpectedly coming across a strange craft and studying it for a while. Then it seemed it suddenly vanished and he was alone and at a different spot on the beach. Moreover he claimed to have lost over an hour of his life. This was followed by recurring dreams of having actually gone into the craft. Such was the situation when the two men first met. Intrigued by this, Mr.Blake, who incidentally did not accept UFOs, became increasingly interested in the case and with Mr.Jones agreement they conducted several hypnotic regressions, until the full experience was revealed.

Mr.Jones felt that by then he had full memory recall of the event and others following it, and at this juncture he resumed the story.

In the first instance he stood looking at the strange machine for a while and as there seemed to be no sign of activity, he made a closer investigation by walking round it. Strangely, he said he felt no fear. He also said there was an extraordinary stillness in the air, other close encounter witnesses have mentioned this. Jones said that the object was about forty feet across and fifteen feet high. It appeared circular and 'resting on some support because the rim was a couple of feet above the ground'. Most puzzling to me was the fact that he insisted the rim of this craft was toroidal in shape 'like a tube wrapped round it' and when the machine departed this began to rotate anti clockwise, but extraordinarily enough it also 'appeared to rotate about its own smaller diameter'!

However before this occurred a sliding type hatch had opened and a helmeted being, clad from head to foot in an overall type suit descended and beckoned to him. He then entered the vehicle in which there were three other similar beings. They seemed to be unaware (or indifferent to) his presence, for if they were aware they betrayed no sign of it, 'they just carried on with whatever they were doing'. The being who conducted him aboard spoke to him in English, though Jones had the impression that this was a mechanical translation. Among other things, he was told they had chosen him because of some particular genetic strain in which they were interested. He was also told that their planet of origin was in a distant stellar system and they had an interest in our world. There were other things they told him, but he felt a strong urge not to discuss this. He was also told there would be other meetings.

Subsequently these occurred when he went alone to a pre-arranged site. The total number of meetings was three. On the last of these they explained that they were taking him on an *extended* journey, but some kind of preparation had to be made for this. He was asked to don a suit much like theirs, with similar headgear and he was warned that for a moment he might experience a little discomfort, but he must not be alarmed by this as no harm would befall him. Next he remembered being seated in some kind of apparatus enclosed within a moveable transparent tube. He also noticed that the others arranged themselves in similar cubicles. Then came the low pitched humming noise he had detected before, while the crew seemed to be intently paying attention to what he thought were various coloured display readouts.

After a few moments of this otherwise silent tableau he suddenly felt a sharp pain in the chest and a tightening feeling in his body, 'something like a cross between an electric shock and anaesthetic', after which the 'crew' detached themselves and him from the cubicles. He was told that the sharp pain was registered at the moment they 'flipped', this being the nearest descriptive term he could identify with.

He had no idea where he was taken or why, but he was given to understand that they were just as curious about him as he was them. He was then subjected to a few biological tests which caused no discomfort. On the return trip the same procedure was reenacted, when again he felt that sharp pain. He was then given to understand that other contacts would take place, but not by these same means. After these experiences Jones life returned to normal, but as he had always been psychically minded he began to direct more time to discussions and meetings concerning the paranormal. This had been the situation when Mr.Blake met him. Mr. Blake then finished the story by telling us that Mr.Jones now held clairvoyant meetings of his own and it might be interesting to attend one of them.

Perhaps it should be reiterated here that the author is of that particular breed who are initially inclined to respond sympathetically, though not gullibly, when first becoming acquainted with information such as this. Within twenty four hours however, the cold analytical engineer in my character takes over and I am inclined to become even more critical than your average sceptic! I wanted to see Mr.Jones again, I had a few questions to ask.

During our first meeting with him I had noticed how at times he seemed to be extremely distant, I also remembered how after that first meeting we had accompanied our visitors to the ferry, whereupon Mr.Jones had become detached and was showing - I thought - an inordinately amount of interest in the boat and its surroundings. I cannot adequately put this feeling into words, save to say, although an otherwise intelligent person, that night in the half light he behaved with almost child like delight, as if he were seeing something unusual for the first time! Then in a moment he was back with us and normality. Along with several of my colleagues we did have another meeting with Mr.Blake in which we heard several recordings made by Jones, but unfortunately I was unable to put my selected questions. However, although it would be some time in the future my relationship with Mr.Jones was not yet over, far from it!

Who can tell what might happen over the next few weeks?

In June 1967 I unexpectedly had a phone call from Mrs.T of Ryde, I.W, who having read *Piece for a Jigsaw* thought I would like to know of some extraordinary meetings they had been having with a trance medium. Some remarkable predictions and healings were taking place and some UFO orientated messages had been obtained, should I be interested, I was cordially invited to attend the next scheduled meeting the following weekend. As I have said earlier, although, out of caution, investigation in this area had tended to disenchant me, on occasion I have felt prompted to respond positively. This was such an occasion, and I said I would be glad to accept. That decision was to prove to be the trigger of one of the most momentous experiences of my lift.

I arrived at Mrs.T's house at about seven o'clock. She was a very pleasant silver haired lady, probably in her late fifties, well spoken and courteous. After a brief exchange of a few local observations she said her other guests had already arrived and were waiting in the lounge for us to join them.We numbered ten in all, four males and six females. They seemed to have been arranged - as is customary in such affairs - in a circle evenly disposed about the medium. In dimmed lighting I was introduced and shook hands with them all, then quite unexpectedly I was surprised by the realization that the medium smiling over to me was Mr.Jones! He had it seemed not only been aware of my taking part in the meeting, but in fact he had positively suggested it. Also for reasons best known to himself, had not revealed until then that he and I had met before.

As I was the last to arrive I was seated almost diametrically opposite to Mr.Jones and now that I was used to the dim lighting, I could not help noticing that same bland smile I was aware of at our earlier meetings. We exchanged a few pleasantries and then Mrs.T proposed we began the session.

Again, as is customary, we were asked to relax and sit in silence for a few minutes,as the medium slipped into a trance state. After a while Mr.Jones, speaking in his normal voice, offered greeting to all present, then began to give certain messages to each one in turn, again as is customary. Apparently this meant something to each sitter, for they responded accordingly. But I was impressed when having arrived at myself I was offered advice on a legal matter, which nobody present could possibly have been aware of. He then spoke of a special event and certain means having been taken to arrange it. The medium said it had something to do with UFOs.

Knowing Mr.Jones claimed UFO experience and his being aware of my own involvement, I could not resist the inevitable, though perhaps uncharitable thought 'here we go'. However after a moment or so of almost inaudible deep breathing, in the dim light his face then appeared to change.When he next spoke it certainly didn't sound the least like Mr.Jones. The voice was difficult to place. It was medium pitch, slightly crisp and with a slight artificial intonation, so slight one had to listen for it, I had a feeling I had heard that voice somewhere before. Later, in hindsight, it reminded me of that other occasion in London in 1954 and the voice in the dark.

I was addressed personally with sufficient reference to some aspects of my work which, among those present, only I could have known. So far it was impressive, though I was not a little embarrassed that I had been singled out for such prolonged engagement. As the minutes ticked by, helped by the polite interest shown by the others, I succeeded in overcoming this initial reaction and concentrated on what was being said. It was of a technical nature going over the heads of the other members at the risk of boring them.

I continued to be impressed. I became certain that the vocabulary and information ran counter to Mr.Jones normal range.

I was told things on which I have often since pondered and was given to understand that the ability of Mr.Jones brain being able to relay this discussion had something to do with his meeting on Hayling Island. The communicant told me the name of the speaker, but it was stressed that this was the nearest that could be offered and I was unable to pronounce it anyway. Whereas, for reasons given later, the nearest I could get to pronouncing the name of the alleged home planet I shall not forget. Phonetically it sounded like XNER. The originating planetary system was left undisclosed. My caution must have been obvious by the manner of questions I asked and I was politely but firmly reminded that such scepticism was a hindrance to the help being offered and I should try to be more magnanimous. Or in more colloquial terms, `I shouldn't look a gift horse in the mouth'.

After farewells and good wishes towards all mankind, about which I sensed a deep sincerity, Mr. Jones slowly returned to himself. Eventually the lighting was raised and cups of tea and biscuits were kindle proffered by our hostess and her husband. This last development had been something new to them and I rather think they expected me to react with cries of `eureka' and great excitement. I just wanted to think about it.

The others present seemed a little let down by this attitude and I apologized, trying my best to explain, but of all those present it was Mr.Jones himself who, now looking a little bit harrowed by the rather long experience, came to my rescue by saying that Mr.Cramp was right in wanting to think more about it. So that most important occasion came to an end, but not before another kind parting remark from him, when still defending me he said,"We must let Mr.Cramp make up his own mind on these matters", then looking meaningfully at me he continued "and after all who can tell what might happen during the next few weeks?" I shall always remember that kindly understanding look and for some reason I could not explain, for a moment I felt humbled to my socks! I left the meeting then and drove home, not really any the wiser, but certainly accompanied by that old feeling of déjà vu. I must emphasize, apart from my wife *I had little opportunity to tell anyone of this event.*

Over the next few weeks things *did* indeed begin to happen. Reports of UFO sightings all over southern England and the Isle of Wight were profuse, some of them taken up by the media, most of them coming back to our Society and therefore me. But of these, the most important case of all was introduced to me personally by the father of the lad who first witnessed it. I must confess that as I began the following investigation all thoughts of the seance and Mr.Jones were farthest from my mind. In fact it was only after several weeks and many miles of trudging over fields and countryside, that I slowly began to recognise a possible connection.Bul long before that belated `awakening' came some of our UFO Society committee members were to be treated to their own personal mini, thought provoking demonstrations.

The first one to be reported to me came via a good friend Tom. By profession he was an aerospace chemical physicist. On a particular night suffering from a nasty cold, he had elected to use the spare bed in a little attic room. He was awakened in the early hours by a peculiar, quite audible, persistent buzzing. Thinking it might be an insect of some kind and not feeling like getting out of bed, he tried to ignore it and get back to sleep. But this proved futile, for no sooner had he begun to doze when a loud rap was heard. By now wide awake Tom heard another and another rap which broke the silence of the room. The night was still with no wind or rain, so now very curious he got out of bed and traced the noise to the small window. He had closed this earlier as he had felt cold. It was still closed but an even closer investigation revealed that the window pane itself was vibrating. As Tom's house was quite near the coast, he was inclined to think that a passing ship was the cause (as has frequently happened to the author). But that could not explain a loud rap. Being a physicist he thought this may have been caused by harmonics of some kind, but suddenly the light from his torch caught something inexplicable. Due to the window being closed the glass had misted over a little and there in one corner were marks which he was certain he had not caused. Closer scrutiny yet revealed that the marks, which were on the inside of the glass, looked like letters, or part of a word which to Tom made no sense at all. They spelled out XNER!

The following evening Tom visited me to discuss the strange phenomenon. I am sure that when I told him my side of the story, the incredulous look on his face was only reflected by that on my own. But even more so, while we were in the middle of this, another of the Society's committee members John, an aerospace structural engineer, came in to tell us of a no less remarkable event which had occurred to him at a similar time that same night! Both these men had flown with the RAF during World War 11 and therefore both were familiar with Morse code. This fact has an important bearing on what now follows.

Like Tom, John had slept soundly until the early hours, when he was unaccountably wakened. He had laid there quietly in the dark listening, for he felt sure he had heard something. He had, and now it came again, he could hear the handle of the door being turned! He sat bolt upright and fumbled for the light switch, waking his wife in the process. The door *was* open, but he felt sure he had closed it as he always did. Puzzled by this and now fully awake, he returned to bed a little uneasy.   Then softly and slowly at first the raps began. Faster and faster, louder and louder, until there could be no doubt about it, they were in Morse. Once more he put on the light, whereupon to his relief, the taps ceased, but not before he had automatically dredged from deep in his mind, a half forgotten habit. It was a word repeated over and over again, it spelt.....XNER! Again it must be emphasized that the foregoing is absolutely true and it is perhaps noteworthy that both these very responsible people were quite independently *reporting this back to me!*

Shimmering telegraph poles!

Douglas W., known to his colleagues as `Duggie', was a technical author. A big man, he stood over six feet tall and cut a very imposing figure in his motor cycle leathers when he arrived at the office. A keen biker, he owned three machines which he doted on and kept in pristine condition in his garage. He was also a UFO sceptic who now and again seemed eager to pronounce loudly and authoritatively on the subject, which I often thought betrayed a deeper interest than he would have cared to admit.

For some weeks past there had been reports of deep holes and crop circles appearing on the island and together with other members of our group I had visited some of these. Then one morning Duggie strode into the office and without much attention to anything else made straight for me. I detected something different about him, he seemed rather intense I thought. Then as if uncertain how to begin he related the following extraordinary occurrence he claimed happened to him after he left the office the previous evening.

Living at Ryde Duggie used to join the main Newport/Ryde road on his way home which at that time in the evening was fairly busy with homeward bound commuters. He was approaching the small village only a few miles ahead where Yvonne R lived (mentioned earlier). To his left were large areas of corn, on the right were a few properties and a service station. Suddenly and so unexpectedly that he was unable to take evasive action, Duggie had spotted something extraordinary, slightly ahead and to his right. Despite the rapidity of the event he said he registered it clearly and would never forget it. A glowing object `the size of a telegraph pole' seemingly floating a foot or so above the road streaked towards him, struck his bike broadside and sent him and the machine crashing into the roadside ditch. More concerned about his beloved bike Duggie struggled to his feet and killed the racing engine. By this time several motorists who witnessed the accident had arrived to assist him, but apart from bruising and a few scratches he was fortunately unhurt. It could of course have been more serious.

I have to confess that as Duggie relayed this story to me I was a little suspicious he may have been pulling my leg. However this soon evaporated when he later showed me the dents and scratches suffered by his bike. He hadn't the slightest idea what the thing was, but apart from the physical shake up and damage, he was clearly stressed by his inability to describe the bizarre phenomenon. He claimed the instant the thing had contacted his bike there had been flashes of light and then it had passed uninterruptedly *right through his machine* `as if it was dematerialized' then carried on into the hedgerow alongside! Any remaining vestige of suspicion Duggie was merely dressing up an ordinary skid or spill was soon dispelled when he came to see me the following day and rather dramatically placed something on my desk. He said while in the process of cleaning the bike he had come across it tightly wedged between the engine cylinder and the frame. It consisted of what looked like a piece of milky white translucent glass measuring several inches across and quite heavy. But when I picked the object up I detected movement, whereupon Duggie turned on my anglepoise light and told me to look closely. The thing was partly filled with oily looking fluid. The surface of the material reminded me of several pieces of meteorite I have, with an irregular nodula texture, as if exposed to high temperature.

By now convinced Duggie had been the participant in an unknown phenomenon, I was grateful when he offered to give me one of the pieces. However this wasn't to be, the material was needed for an insurance claim. Perhaps not surprisingly he never got it back.

Quite a few years later I came across an article concerning a farmer on whose fields there had been appearing crop circles. He had been driving his tractor down a narrow lane approaching a gate in one of his fields, when a copy book repetition of this phenomenon occurred. He also described it as like a white floating telegraph pole, which shot out of the gate narrowly missing his tractor and *through* the *closed* gate on the opposite side of the lane! This will be the first time I have publicly disclosed the incident which occurred to the UFO sceptical Duggie W, but regretfully I have since lost my notes concerning this latter corroborative affair, but perhaps it will be recognized by the reader.

Whippingham and a footnote

The major episode which was then to follow eventually appeared in the May/June 1968 issue of the world wide publication *Flying Saucer Review* and it is suggested that those who haven't already read this will be more appropriately informed if they can do so.

As a footnote to this chapter I should add I am aware of the fact that some might be inclined to think that having by nature been so closely related to paranormal effects for much of my life, together with a natural bent towards scientific enquiry, has been a recipe for possible conflict. However this is the antithesis of the truth, for to me there can be no distinction. In other words, as stated elsewhere, I regard the so-called `paranormal' as an intrinsic part of the so-called `normal'. This does not mean to imply that I am unaware of the fact that some might find it bizarre that an extremely advanced technological race would use a human medium for communication. I too have passed through that phase. But when you think about it, such a process would be even more advanced and by far more convenient than our present day radios. Even the most elementary acquaintance with the function of the brain bears testimony of that.

To which I can hear someone say `even so....but taps on the wall?' May I reply by offering a fundamental analogy. If you wanted to communicate with a friend who was occupying an otherwise empty room, without the assistance of artificial aids - mobile phones etc. - what option would you have left other than by direct communication via the most sophisticated radio of them all, the human brain. That is, *if you knew how !* Moreover if your friend's `receiver' was an early model not yet tuned to the required frequency range, you might find an intermediate solution would be to select just one band or idea, which your friend's early model might pick up and he would get the audio *impression* of a rap.

If this would imply a complete reversal of the normal function between the body and the brain, then it is not entirely without foundation. For example, it is well known that subjects in a hypnotic trance have exhibited bodily effects when a strong suggestion has been implanted. Indeed on one occasion the author's mother when a young girl received a substantial burn to her finger simply because she inadvertantly touched a cold stove which *for that precise instant* she had been utterly convinced was alight. Such a phenomenon is of course commonly known to medics as auto suggestion.

Not least, it is worth considering that the direct mind to mind process (telepathy) is probably a natural God given inherently private affair, not yet available to the twitching ears of most of us and therefore unlike our comparatively primitive radio systems, it remains isolated from interception or `bugging'. Also in this context some Ufologists, and many sceptics, have treated with suspicion witnesses claims that attempts to photograph a grounded UFO failed due to the fact that it appeared to vanish, or as in my own case, there was a failure of photographic equipment. However this suspicion is somewhat negated if one considers the extreme lengths to which our own authorities go in order to protect highly secret projects. With rockets and aircraft at one end of the spectrum to the family car at the other, let alone the technology behind a highly sophisticated inter spatial vehicle we commonly call a UFO. I put it to the reader, there is nothing mysterious nor sinister about this kind of `sensitivity'. After all. should our respective positions be reversed, would we behave any different? The honest answer has to be No, and for very good reasons when you think about it.

FLYING SAUCER REVIEW

Vol. 14, No. 3   MAY/JUNE, 1968   14th Year of Publication

Part of the extensive damage to one of several Isle of
Wight barley fields. Did a **UFO** pass this way? See . . .

THE WHIPPINGHAM GROUND
EFFECTS

PAGE 3

FIVE SHILLINGS

*Cover of the popular Flying Saucer Review magazine which carried the article on the
Whippingham ground effects*

THE WHIPPINGHAM GROUND EFFECTS

WAS THE DAMAGE TO CROPS CAUSED BY A UFO?

By LEONARD G. CRAMP

During the summer of 1967 the flight of an unusual aerial object was observed by schoolboys in England's South Coast holiday island, the Isle of Wight. Our contributor, well known for his two books, *Space, Gravity and the Flying Saucer* and *Piece for a Jigsaw,* has had many contributions in the pages of FLYING SAUCER REVIEW, including one in our very first issue in January 1955. The investigation which he and his friends made of this 1967 incident was so meticulous that we are delighted to present the report for our readers.

AT a quarter to nine on a fine, cloudless morning last year (July 10, 1967), two pupils of the Whippingham Primary School, near Newport, Isle of Wight, were lining up with their fellows to enter school when a stationary object in the northern sky caught the attention of one of the boys. The object was "milky-white" and "cloud-like" and looked like a disc with a bi-convex section. It may or may not be significant that immediately prior to the sighting the boy's eye had been attracted by falling ash, drifting in a *westerly* direction across the sky. Apparently ash does occasionally drift across to the school from a works situated farther to the west, but then the ash is carried *eastwards* by a westerly wind. On the morning in question there was no wind, and "there seemed to be silvery sparks around the large pieces of ash". It was while looking at this that the UFO was first seen hovering near the distant B.H.C. test tank research establishment.

When the boys came out into the playground at 1030, the boy who had spotted the original object looked in the direction in which he had first seen it at 0845, but saw nothing. Then his companion spotted the same (or a similar) object further to the west. About a dozen boys watched as the UFO moved in a westerly direction. Its size was estimated to be "larger than a bus". During this time the UFO was descending and was "fluttering down like something out of control". This motion continued until the UFO was at approximately 2 degrees elevation; then it seemed to "correct itself" and began to climb. The westerly traverse between the first and second sightings was calculated to be approximately 30 degrees. The UFO was then lost to view behind a line of trees.

On his way home that evening on top of a bus, one witness saw marks in the barley field (Site A) which is bounded by the Newport-East Cowes road and adjoins the school playground.

Investigations of this site revealed large areas (up to 6 yds. wide) of damage, in the form of depressed and flattened stalks, which made an almost completely circular pattern. The damage had a very mechanical appearance in a vortex pattern, sometimes clockwise and sometimes anti-clockwise, but predominantly clockwise. The centres of some of the vortices had tufts with broken stalks and others had nothing—obviously

the roots and stalks had disappeared completely. In these areas the heads of corn had been denuded and looked (to quote one farmer) "as if they had been thrashed". From the rim of several of these vortices there were "lanes", about 1ft. in width, which began in the barley that had been pressed down outwardly. These lanes tapered to an end in the midst of untouched barley.

At first the investigators presumed that the damage was restricted to the area near the school, but further investigation revealed that the marks in the Site "A" field continued, in a diminishing pattern, in a northerly direction parallel to and 12 yds. from the hedge lining the Cowes-Newport road. The marks on the Northern and Western boundaries of Site "A" were discovered at a later date when a study was made of aerial photographs.

The barley field denoted as Site "B" has a piggery on its boundary, and the damage here was discovered to be in the form of a continuous trough some 3 to 4 yds. in width and 88 yds. in length. The trough runs close to —and parallel with—the access path to the piggery. There is a strong resemblance to the general characteristics at Site "A". The undamaged heads of corn on the northern side of the trough were in a uniform line towards the east, as though the barley had been swept by a broad broom. The damage discovered on this site corresponds exactly to the 30-degree traverse indicated by the boys. The westerly end of the trough thins out and skirts round a small derelict shed. This thinning out coincides with the point at which the object was said to have been climbing. This completely supports the suggestion that, if an aerial object produces effects on the ground because it is at a low altitude, then those effects must diminish as the object gains height.

In the "eye" of one of the whorls a 6in. cube section of concrete, weighing a few pounds, was found resting on some of the stalks but covered by top stalks. Although the investigators tried to identify the piece with others on a nearby disused gun-site, they could not satisfy themselves that it had come from there, although this is the likeliest possibility.

A further significant point is that the body of a wood-pigeon had been seen at the precise point at which the witnesses stated the object was first seen hovering. This

3

Extract from The Flying Saucer Review article on the Whippingham UFO ground effect 1968.

View from school showing UFO flight path beyond the line of trees.

The corn damage con -tinuing in Site `A' towards the distant B.H.C. Test Tanks.

Aerial view of Site `A' looking S.W. showing school boundary at top left.

The corn damage in these photographs was only discovered after the school children had indicated the flight path of the UFO shown in the top photograph. The radial lines in the centre of the lower picture are land drains.

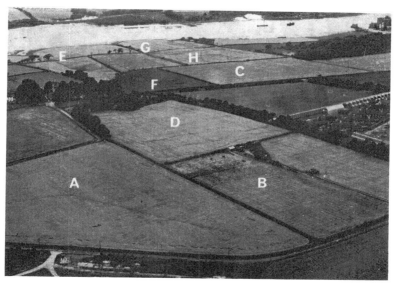

Aerial view looking towards the River Medina

Typical rotary damage on Site 'C' in which the last
remains of the little wood pigeon was found. This photo shows
one wing, the many other feathers are camouflaged among the
barley. Here again the broken and thrust down effect is most
evident

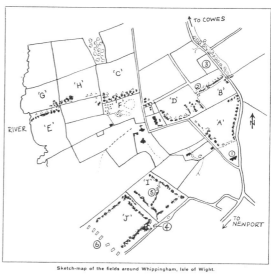

Sketch-map of the fields around Whippingham, Isle of Wight.

Key: 1. The School   2. The piggery   3. British Hovercraft Corporation test tanks
        4. Derelict harvester   5. H.T. poles   6. Caravan site

Note: With the exception of 'F,' the lettered "sites," and also that lying between 'H' and 'F,' were all barley fields.
It was in these that the damage was discovered. In the sketch areas of damage are marked by the dark, curly pattern,
and should not be confused with the conventional signs for trees elsewhere. The positions of the stones are denoted
by crosses in sites 'B' and 'F,' and the approximate areas where pigeon feathers were found are indicated by small
triangles.

*Aerial photograph and map shows the continued flight path of the reconnoitering
UFO including its continuation on the other side of the river Medina at top,
testified by several independent witnesses.*

5
UFO Interlude

1967 - 1994

After the events which culminated in the Whippinghan case, UFO sightings on the Isle of Wight began to diminish and of these most were typically of the LITS (lights in the sky) variety. Personally I felt something like a mountain climber desperately wanting to find a ledge on which to rest. Also I have to admit, that I found it hard to resist translating some of these `near to home' phenomena in terms of my fascination for gravity and energy. Sometimes I found myself yielding to the thought that, had someone else in similar circumstances told me such events had overtaken *them,* I would have been surprised if they *had not* suspected a possible correlation. In a somewhat vague kind of way it was as if someone had said, `by now you should have got the message, now there is some work to do!'

To what degree I had been influenced by all this I truly cannot tell, save to add that for the next few years, apart from responding to the odd invitation to lecture on the subject, and the odd case now and then, I gave public UFO investigative work a miss. Even though this involvement was sporadic and spread over time, I suppose in a way I had absorbed enough. Meanwhile I had made some exciting inroads in my ongoing work towards levitation and more down to earth efforts in alternative solutions for conventional VTOL (Vertical take off and landing) aircraft. The latter being typified by a concept I labeled the `Hoverplane', shown later on. Not least the then I.W. Council had become actively interested in another of my schemes called `Vistarama', also discussed later.

These developments were very appealing as a future for our family, so after much debate I decided to leave The British Hovercraft Corporation and venture out on my own.

Thus in 1971 I was independent, and by some standards I had previously endured, comfortably furnished with one room in the house as drawing office and large garage as a workshop. My wife had a job, I was freelance for Matchbox Toys, the family were growing up and leaving home, so for the next two and a half years we soldiered on.

By 1973 it was obvious for the aircraft research work - and us - to survive, I would have to seek sponsorship funding. Also the large models we were building for `Vistarama' demanded more and more space. We had to find alternative accomodation with a workshop, but where and how? Once again came that remarkable helping hand. After looking at dozens of properties in which `out buildings' had been quoted (usually proving to be little more than corrugated galvanised iron sheds with two inch broken cement floors etc.) we literally homed in on a very nice small bungalow set in woodland on the coastal cliff. Apart from stating the property was about three miles outside Yarmouth, the short

Aerial view of Green Acres (later to be labeled Len & Irene's Ranch House by Desmond Norman) as it was when we first saw it in 1973. The 'shed' in which the majority of the research has been done lies in the foreground.

24 years later, front face of the workshop sporting reclaimed windows, repatched cladding, coats of paint etc., all achieved by a good deal of hard work by the family.

provisional 'watch this space' type of advertisement, did not disclose the actual address. But it did say in the grounds there was 'a hutment', which I interpreted as being of the forestated shed variety.

However, acting on a hunch I placed an ordnance map on the table and with a pair of compasses, swung a scale three miles radius around Yarmouth in which I thought was the most likely area. It was quite simple, anyone could have done it; three miles to the east which was country and near the cliff face.

As we like a drive in the country to locate out of the way roads and lanes just for the fun of it, we took the car and went to find the 'guestimated' spot. Then turned off into one such unmade lane. After about a mile we turned into an even smaller lane, at the end of which, set among the trees was this delightful chalet bungalow set off by about three quarter of an acre of grassed area. Prompted by that now dominant urge I got out of the car to look, accompanied by a reprimand from my wife. After standing enraptured for a few moments, I turned and saw the most beautiful view across the Island, certainly if that place had been for sale, and if it was within our bracket, I would love to have it. workshop or not. But then my reverie was interrupted by a call from my wife who was then gesturing towards the house. The owner having spotted us had come out to investigate, so not wishing to inconvenience him I moved towards the car. At this the man called and beckoned me to join him, which, with apologies I did. Far from being annoyed he was delighted for he said he had only put the property on the market that week and as the provisional advertisement came out that Friday, he hadn't expected to have a response yet, would my wife and I like to come in? We were dumbfounded, not only was the property for sale, but it was *the* one we had seen advertised, we were the first to see it and it was within our means. But the extraordinary 'bonus' was yet to come.

My wife and I were introduced to his wife, then he took us around the house while she kindly prepared tea. In the garden to the rear of the house we came across the typical galvanized roofed hutment. True it was little more than a disused stable, but I liked the property sufficiently to put up with that. Also the large grassed area and woodland garden of two acres together with the surrounding paddocks would be admirable for our model aircraft tests.

Sipping tea and admiring the view from the front of the house, I noticed the nearest neighbouring property situated toward the corner boundary about one hundred yards away. Like a huge barn, it looked to be about one hundred feet long by at least forty feet wide, "who does that belong to?" I asked casually pointing towards the building, "Oh that" came the astonishing reply, "goes with the house". I had found, or been led to my workshop!

As the owners had only lived in the house for just over a year, they had rarely gone to look inside the building, in fact they had difficulty in finding the key. I left the three of them, went over to the building and stood alone in its vast emptiness and gave thanks in gratitude for this remarkable contribution to my work! With the help of my two sons I was to spend the next fifteen years working there in all weathers, at all hours, often late into the night, in fact exactly as predicted by the medium's message in London in 1954! We suffered all manner of disappointments, hardship and worries, yes and we had the most wonderful success, I would not have exchanged it for my earlier secure sojourn in the establishment for anything.

A bizarre cliff top encounter

As stated, during this intervening time, UFO reports of the Nocturnal lights variety continued to be reported, but there were two exceptional cases which came my way.

The first incident occurred in the summer of 1975 and as it involved two people who wish to remain anonymous I shall call them Phil and Joan. The weather had been hot and humid that day so they decided to park their car and take a late evening walk along the cliff tops on the southern coast of the island. The night air was already cooling by a gentle breeze and there were only a few scudding clouds pleasantly lit by the moonlight. They had walked further than intended and having found a cliff top seat at a vantage point, they sat down to admire the scene and rest for a while. By then the night was calm and still and the silence was only broken by the gentle swish of the incoming waves on the beach some eighty feet below them.

Out at sea they could see the lights from several fishing boats, presumably at anchor for they did not move. But Phil had spotted one larger, brighter light which appeared to be growing still larger. Realizing that whatever vessel the light was on was coming towards them rather fast, he had pointed it out to Joan with the remark that perhaps it was a power boat or even a BHC hovercraft on night trials. However it now became apparent that the light was in fact formed by several smaller lights, which as they came nearer resolved themselves into several horizontally spaced orange lights surrounded by a single red light, the whole being diffused by a surrounding glow. Phil was still inclined toward the hovercraft explanation, but he became puzzled by the fact that there was no noise! Within moments they could make out the vague dark silhouette of the craft which carried the lights, but by now Phil realized that whatever it was, this thing was some distance above the sea and coming straight at them!

Now standing and in some alarm they had watched fascinated as the object shot rapidly nearer and finally became lost to their sight somewhere below the cliff edge. They held their breath expecting the dreadful noise of a collision as the thing impacted into the near vertical cliff face, but there was nothing, save for the fact that somewhere down there on the beach something was still glowing, for they could see the cliff top - only some twenty yards distant - silhouetted against the glare.

Still thinking that there had been some dreadful accident and despite Joan begging him for caution, Phil edged toward the cliff top, then froze in his tracks, for above the background noise of the surf there was a strange deep throbbing hum, as Phil said "something like an electric motor". Later both witnesses told me there was also a smell like `rotten eggs'.

Suddenly the glare and the noise faded, leaving two bemused and rather scared people in comparative darkness. Phil had gingerly traced his way back to Joan and the two sat on the seat quite uncertain as to what to do. Phil said they had better hurry back to the nearest phone and inform the police or coastguards of the occurrence, but for them the episode was only just beginning!

By now, more accustomed to the gloom, they were in the process of rising to make the phone call, when Joan gripped Phil's arm and pointed in terror towards the cliff edge, she had "seen something, something moving". Despite the uncertainties and against the pleas from Joan, Phil again edged his way forward. Joan's eyesight must have been more acute in the dark, for at first Phil couldn't see anything and called out as much. Then once more he froze in his tracks, for now he too could see a dark dome shaped mass just rising above the cliff edge. He had hesitated long enough to detect some kind of dark protuberance from which came a very dull phosphorescent green light. At that point he had turned and started running back towards Joan, who by now had anticipated him and grasping hands the pair raced and stumbled over rabbit holes towards a phone and safety.

The next thing either of them knew was that strangely they were sitting on the grass near the seat, rather cold, but now all fear of the episode had left them. Nevertheless,still puzzled, they had returned to the car park, got into the car, lit a cigarette and sat trying to decide what to do next, for by now they both felt something very unusual had happened. Phil decided he would inform the coastguard that there may have been an accident and leave it at that. He was half way out of the car when he realized, according to his watch, it was now nearly 2 a.m. which was impossible, for the return walk and the time of the strange event, could have taken no more than one and a quarter hours, and they had begun their walk around 9.35 p.m. So his watch must be faulty. But a quick check with Joan's watch showed nearly 2 a.m., they had been asleep or lost nearly three hours of their lives!

Short of alerting the coastguards, the couple thought there was little more they could do and resolved to keep the strange experience to themselves. Which they did until several weeks later when learning of my interest in such matters, motivated Phil to contact me, no doubt hoping I might be able to throw some light on the bizarre affair. When Phil first visited me he was still in a state of shock, trying to control the odd shiver as he described the extraordinary train of events. He said that shortly after this experience, Joan's home had been plagued with poltergeist activity.One pound bags of sugar sailed through the air, crockery was thrown about without damage, footsteps and voices were heard, her cat, normally a placid animal, behaved strangely as if petrified etc. I eventually tried to reassure Joan and after a few weeks the phenomenon did in fact begin to subside with only the odd recurrence, which finally stopped altogether. Some time later Phil had a couple of experiences of bi-location. The first when he was *seen and spoken to* at a party, the second while he was talking to someone in the High street. On both occasions he had witnesses to the fact that he was at his own home! Strange dreams and the loss of time on the cliff disturbed him to a point where a hypnotic regression was attempted, but if the hypnotherapist learned anything of significance I was never able to determine, for sadly not long after, he died of cancer. Today Phil is a dedicated but low key Ufologist and accepts his experience with a quiet reserve.

Glimpses of the future?

This other case reported to me occurred on January 14th 1976 witnessed by a Trinity House officer Mr.Rod R.of Ryde, I.W. As a trained observer his descriptive ability, together with that of his wife, was so profuse that with their cooperation I was able to produce a `photo fit' likeness of the object they had seen.

At forty minutes after midnight Rod R. went to lock the back door of his house before retiring , when a large stationary light in the sky attracted his attention. Due to low overcast there was no moon or stars visible and he estimated the apparent size of the light as about the size of the full moon and at first a small portion of it was hidden by low cloud. Rod said he leaned against the wall and watched the light for about a minute, then it became extremely bright and started to move very slowly towards him, travelling from west to east. He then went into the house and called his wife to join him. When they arrived outside the light was still slowly approaching the house. As it did so it became apparent that beneath the light there was a large dark silhouetted object. By the time it was overhead they could see the shape completely, it was rectangular and Rod said that due to the cloud height he now estimated the height of the object to be no more than three hundred feet and the craft to be about one hundred feet long and fifty feet wide. He said, by then the light was more of a glow, seemed to emanate from the object and was roughly the same size. There were no other lights or markings and there was no noise.

The couple then had to move to the other side of the house in order to see what the UFO would do next and they were just in time to see it accelerating away and upward at a tremendous rate before it disappeared into the cloud cover. The witnesses stated that it was a very still night with no wind or traffic noise.

The following afternoon Rod rang the Royal Air Force base on Thorney Island. Who said they would telephone the Ministry of Defence and they would probably contact him direct, which they did within five minutes. Having been given the details, the Ministry official said Rod would be hearing from them again. Eventually he did receive the usual letter which after thanking him stated that the MOD would be examining the report 'to see if there are any defence implications'...and 'not to advise you of the probable identity of the object seen etc.etc.

Shortly after, the sighting was corroborated by the duty Met., Officer of RAF Bracketts, who described the craft and circumstances as to wind and cloud height, exactly as Rod himself had done. But within the context of this present work, the intriguing thing is that although Rod knew of my interest in UFOs before he contacted me, he had no way of knowing that at that precise time I was working on my Hoverplane project and in particular the version on which I was working had a *rectangular planform.* Of course UFOs as such come in a variety of shapes, but they are predominantly of circular, rugby ball, cigar or even delta configuration, hardly ever rectangular.Thus it was as I changed my rough sketches to suit Rod's prompting, I found myself sketching my own Hoverplane configuration, which was exactly a two to one length, width ratio and with a radiused *transparent* upper body. This is perhaps of only slight interest to note, but as I have shown, designers of today very often return unwittingly from sleep time states having had glimpses of things yet to come.

Here again these cases have never before been publicized and I quote them here not only because I can vouch for their authenticity, but additionally because of the paranormal and time slip phenomena which accompany them.

Déjà vu

One afternoon in 1979, (several years after I had left the British Hovercraft Corp.) I received a telephone call from a young man employed there. He said they had been entertaining some Russian visitors, both civilian and military who *during lunch* had been asking inquisitive questions about my book 'Piece for a Jigsaw', hence the phone call. Had I any remaining copies available? If so he would send someone to collect them. The following day I received yet another call. The Russian visitors were catching the next boat from Ryde and if possible they would like to have a few extra copies of the book.

The Russians got their additional copies of 'Piece for a Jigsaw' and made their boat connection. The young man at the office was pleased and in a quizzical kind of way so was I. Yet - as shown in Chapter 3 - how could I even begin to tell him of the strange déjà vu experience in which he had just been an unwitting player?

How do we reconcile
a paradox like that?

I succeeded in my bid for sponsorship, helped form a company and appeared on the well known BBC program, 'Tomorrow's World'. At the time of writing the Hoverplane concept is still attracting attention in different parts of the world. But of most importance, I finally developed a field propulsion theory to a degree which inherently suggests the availability of so-called free energy for the not too distant future. It was in my workshop and many, many hours of searching that I extended Antony Avenel's Unity of Creation Theory, (as shown in The Cosmic Matrix) to explain that indeed magnetic attraction and repulsion are *identical* phenomenon as is gravity itself. How we might profitably adapt this principle for all our energy requirements. But it was also in that workshop that I would come face to face with the stark realization, that on the one hand we *must* find such alternative energy and propulsion systems soon, while on the other hand, if we do, it implies that any owner of a new wheeless car of the future, inherently owns a vehicle which is potentially capable of going into space! Times without number I asked myself 'In terms of some factions in today's society, how do we realistically reconcile a paradox like that?' It has to be accepted that some motorists can be a menace on the roads. But I could not equate them having a wheeless vehicle which would enable them to travel silently over anyone's garden fence, day or night!

All engineering is a logical sequence and in that discipline if you are unable to find a logical solution to a problem, you have little alternative but to compromise. That is the way I felt when I interrupted the more empirical side of this work and put my reluctant conclusions down in a book. The arguments for this decision being adequately supported by the readily available statistics elsewhere. But currently and perhaps not surprisingly, nobody seems to *want* to know about such vital facts.

For rather obvious reasons I had tentatively entitled this proposed new book Pandora's Box, but then out of due regard for Antony Avenel it was changed to the recently published The Cosmic Matrix with much of the 'unpalatable' statistics omitted.

An objection but from whom?

During the summer of 1989, Timothy Good had kindly offered to hand the manuscript for Pandora's Box to his publishers Sidgwick and Jackson, which in due course introduced me to the editor Carey Smith. Whilst in the process of evaluating this work Carey asked if I would consider a reprint of Piece for a Jigsaw as Timothy considered it to be a 'classic'. Of course after all the wasted years I was delighted. But then I was astonished when she went on to add that based on their successful annual Year Book of Astronomy edited by Patrick Moore they were considering the publication of a similar annual series entitled The UFO Report, and would I consider being the editor. I was extremely grateful but the implied work load for *three* book projects together with my existing work flashed awesomely through my mind. I thanked Carey and said I would think about it overnight and let her know. The following day I phoned Carey and told her that despite certain misgivings I had decided this was an offer we could not refuse, and without much ado went on to discuss some ideas about the proposed UFO annual format. Within a few days I had completed a small rough booklet with a jacket design and proposed layout and posted it to Carey. Later when I telephoned her she seemed very impressed and chuckled over the little saucer picture I had scribbled on the jacket.

It was during this discussion that Carey said she was intrigued about the fate of Space, Gravity and the Flying Saucer, and Piece for a Jigsaw, especially when I told her of the repetitive bizarre background to both these books. And rather foolishly perhaps I went on to point out that such a fate was hardly likely to be reenacted with such a large company as theirs etc. The following afternoon I was preparing a few lines for Pandora's Box and among other items on the table before me was an article about the MIB (Men in Black), when the telephone rang. Having received only silence I was about to replace the receiver thinking I had a dead line until I thought I detected a very faint voice on the other end. To make sure I strained to listen carefully then realized there was indeed someone speaking, no doubt due to interference, in a fairly high pitched tone. Believing I had interrupted a crossed line I was about to replace the receiver again when I realized that whoever it was this person was rather angry, I had unintentionally got hooked on to a private blazing row. But my brief hesitancy revealed a more directive feel that somehow that person *was* angry *with me*!

By this time I was listening intently rooted to the spot, not wanting to miss what this was all about. The final words were distinguishable and couldn't be mistaken viz..... "''ex directory....as sure as hell, good-bye!''. If I heard the rest of it subliminally I don't know, but I suddenly felt very guilty and had little doubt that I *had* unintentionally offended someone. If so I could hardly blame them, for I *had* rather stupidly made a challenging remark. Asked by my wife who it was on the phone, I made some excuse about wrong number, went back into the room and there in the open pages on MIBs before me was a reference concerning the uselessness of some Ufologists trying to avoid sinister phone calls by going *ex directory!*

During the rest of the day I gave the matter a great deal of thought. I could do without this sort of thing. I had Hoverplanes, gravity and energy and all kinds of other things on my plate, surely that was sufficient! Carey Smith was disappointed when I telephoned apologizing for any inconvenience in canceling our agreement, but now was the need to find someone else to take over the work on the UFO Report. I thought the least I could do was to help her in this respect, and despite the fact that we both knew Timothy Good was heavily committed to his new book, I considered he would be admirably suited to the job as editor for the new annual and proposed as much, which he kindly accepted.

Reception loud & clear

In addition to the above mentioned work, at that time our family having long since married and left home, my wife and I had reluctantly decided to move in order to live with my elder sister who was now rather infirm. In order to do this I designed an extension to her property. But I still needed a drawing office come study and the best place for this was the attic space. It took a lot of doing and once again we all took a share at building.

Eventually we moved in and tried to pick up the threads of our lives and in particular I was completing a revised version of Pandora's Box from which I had long since deleted the entire UFO section.

Alone in my study late one evening I was reviewing the omitted section with some reluctance thinking, what a waste, surely there was justification for replacing it at this late stage? I was very tired and undecided what to do, save that I wanted to do the right thing for all concerned.

Before me was a plain pad sheet, and pen in hand I was pondering and doodling as one often does, when my reverie was suddenly interrupted by the gentlest of contact taps (which I have never abandoned nor taken for granted and consulted only when support was desperately required). Half asleep I realized I had doodled the large words YES and NO and now I had just received two confirming taps for NO. Instantly wide awake but a little unsure, had I got it right? I was instantly assured I had. Before this occasion I had rather naturally speculated about a possible link between this paranormal phenomenon and UFOs. Now that was about to be rectified in no uncertain terms.

My drawing board, now doubled up as a desk, was situated beneath the sloping attic roof. Ahead of me we had fitted a large `Velux' window, the fresh air vent of which was open permitting the occasional sound of a passing car in our otherwise quiet cul-de-sac. Suddenly, seemingly immediately above my head, came an extremely loud report as in an explosion. This was immediately followed by a second which shook not only me but the entire roof causing a veritable cascade of loose tile chippings between the roof battens. Within seconds my wife was calling out "what the blazes was that?"

Situated as we were only a couple of miles outside Yarmouth and the Lifeboat crew living there, we were used to the maroon being fired in times of emergency, one rocket for stand by and two for action, but these were distant sounding reports, nothing like our very localized example. So neither the maroon, sonic boom nor even a car engine backfire could logically explain such a potent noise. Next I called our neighbour asking him if he had heard anything. Indeed he had, and in no uncertain terms he said he had just stepped out of his car when he was shaken by two loud explosions, which seemed to emanate over our roof causing him to look up in that direction, however he saw nothing.

The following day I looked for damage on the roof but found none, then went up to the attic study and sat down to ponder and there on the desk before me on an otherwise empty pad were those two scribbled words; Yes .....No! I took the hint, put the relevant UFO section back into its file and got on with the rest of the manuscript which was to become The Cosmic Matrix as published in July 1999, which now with the benefit of hindsight has *proved to be the correct thing to do.*

The law of tuning in action

Of course there have been many UFO researchers who have had similar experiences to the author and some have contributed written testimony of that. Among these are some well known UFO pioneers, but very few professional scientists. By no means do I imply by this that there haven't been scientists who have - and do - experience precognitive dreams and/or UFO related phenomena, indeed I have known as friends some who have. But most professionally employed scientists are reluctant to declare this publicly, least it irrevocably damages their egos and/or their reputations, to say nothing about their credibility with their employers. However, as I trust has been shown, my interest in space flight , gravity, UFOs and my professional career developed side by side, and those who employed me were aware of this. So for me the above 'reputation risk' posed no real threat. Therefore as I have stated elsewhere, with that exception, I regard my experience as nothing special. But the whole object of this book will be to no avail unless the reader accepts my professional word, that all that is offered here as supportive evidence to help reconcile some of the apparent diversification among the UFO phenomenon, is absolutely true.

As I stressed at the beginning, I am not inclined to the belief that these happenings in my life, both personal or UFO related, are in any sense odd, rather it is my opinion that we all have an innate ability to achieve just about anything we strive for. In my professional life I have seen this substantiated again and again, in fact everything we have today is the result of continued dogged effort to realize a dream on someone's part. Igor Sikorsky, 'father' of the helicopter, was derided when he flew his first clumsy machine which by all the rules of the day should never have caught on. But it did, simply because he wouldn't stop trying. And this miracle of engineering can be reflected countless times throughout the whole spectrum of our present technology. In other words, if we aim for something stubbornly enough, we can reach it. In my own searching I have seen this borne out many times and not least it seems in the enigmatic domain of the UFOs. So the ultimate realization is not a fortuitous accident, but simply the law of *tuning* in action. In my own case I have wanted to know about such matters, sometimes as if I were remembering them. Thus it is that there is a reciprocal effect, we draw such experiences towards ourselves and profit from the sign posts we meet on the way. In this manner my foregoing precognitive dreams have indicated strongly to me some otherwise hidden truths.

In a quest for evidence could
anyone have asked for more?

In summary of this admittedly extraordinary catalogue of events, I am prompted to stress that during my extended UFO participation I have constantly felt, and at public lectures on the subject have stated, that I have tended to have reservations where certain observations are concerned and that I feel more comfortable and reassured if I can view positive, *physical* UFO after effects, or there is shared participation and more so if *an animal takes part in the incident.* So now, while I have no wish to compete with the cynical UFO sceptic by appearing guilty of presumptuous bias, truly, in a quest for evidence, could anyone have asked for more?

Predictably there will be those who will feel much of the foregoing is purely circumstantial. In which case as an aerospace technician I can assure them that many an aircraft design factor has been based on far lower probability statistics than this. While in a court of law neither would I favour anyone's chances in trying to persuade a jury that all the foregoing experiences were merely accidental! In truth they represent only a small part of the whole. But of no less importance is the fact that this present contribution, to what is after all one of the most important issues of our time, may offer one more useful piece for this remarkable puzzle. Not least, in view of the fact that someone, somewhere it seems, has gone to an awful lot of trouble in my belated exciting `education', I feel the least I can do is share it with somebody else.

Some sceptics might be tempted to consider that I have fabricated the foregoing or even `engineered' it. Indeed, short of being introduced to all those who were involved - and many of them have sadly since died - what evidence is there to show otherwise? After all they may argue, I may have written this book solely for monetary gain. In which case they would be partly right. *Anybody* who writes a book has money in mind! However in my case it would be nice to repay my wife back a little for the life time she went without holidays, clothes and evenings sitting at home with the children, typing my manuscripts and letters, while I went all over the country investigating, and often as not,giving unpaid public lectures on UFOs. To say nothing of the expensive lost earnings and mortgage incurred in private gravity research, designing, building and flying unsponsored test models that in industry would have cost hundreds of thousands of pounds. Yes, unashamedly I want to make money, and I think most people would agree, it's about time I did!

But it is of no less importance that people should be made aware of what is going on, both in our back yard and that due to overspill of some of our neighbours. There is a *real* message in my books, the foregoing is merely a credibility introduction to that message, and it is addressed to people of all political and religious persuasions. Not wishing to moralize, I am prompted by some of the information that it concerns us all, and I trust I will be forgiven for relaying it here; `For God's sake let us stop behaving like children squabbling over-what in the final analysis amounts to petty global and entrenched vested interest issues-and get it firmly uppermost in our minds that unless we find a meaningful alternative to energy and way of living *very* soon, there wont be a world over which to squabble!'

A belated postscript

Recently I was given a little book entitled UFOs over Hampshire and the Isle of Wight by Robert Price. In summary to Chapter 4 (The Isle of Wight enigma) commenting on much of the foregoing material and more, it is interesting to note Price said " Unidentified flying objects know no boundaries - they are a global phenomena. Now and again a sighting of exceptional quality will attract world wide attention and another link will have been added to the long chain of UFO `knowledge'. When the chain is complete and the mystery solved, at least one of those links will have been forged on the Isle of Wight!"

PART THREE

Precognitive Excursions
of the Aeronautical kind

6
Jet Propulsion to Skimmers

142 - 1994 Miniature Gas Turbines
ATF 52 years

As from now on we shall be talking the language of `nuts & bolts' technology, it is as well to remember the three pronged aspect of this book i.e.to offer a little insight into the background of the so-called paranormal and in particular show examples of a precognitive technological nature. To introduce the reader to the world of aviation and thereby illustrate how easy it sometimes is for people to make erroneous deductions, notably concerning UFOs. And lastly offer a degree of credibility support for the observations and disclosures this author has made elsewhere and in particular The Cosmic Matrix.

As previously stated, due to a constantly interrupted and varied work load, it is extremely difficult to maintain chronological order of the various works shown. However the dates quoted give an accurate measure of any advanced time factor involved, for convenience, designated ATF in years.

Welcome then to the following brief introduction to the practical side of the author's early `education'.

Small beginnings towards UFOs,
giant sized jets and gravity

As discussed in Chapter 1, in 1942, having served my `apprenticeship' with Taylor Woodrow, the Ministry of Employment directed me to the De Havilland Aircraft Co., at Hatfield, Hertfordshire as a technical illustrator. At that time De Havilland were building the famous WW11 plywood and balsa constructed Mosquito and Hornet aircraft series. Therefore there were as many cabinet makers and furniture designers to be found there as there were panel beaters and structural engineers etc. In hindsight it may not have been strange that the M of E should have sent me to De Havilland's as they did. As has been shown, it would prove to be the beginning of an aerospace adventure for the rest of my life.

Aircraft, gas turbines &
panic alert at Hatfield

I immediately became fascinated by the first turbojet engine as something vaguely familiar I had imagined as a youngster at school. Constantly being away from home no doubt helped fire an enthusiasm for gas turbines which occupied a great deal of my spare time and gave wings to an intense wish to design and build a miniature version for a model aircraft. This was an extremely difficult thing to attempt as I was in lodgings at the time. No doubt recognizing my enthusiasm the chief experimental engineer allowed me to use a friend's work bench in the corner of the vast hangar which housed the first, then top secret, Vampire jet. I well remember the day when in the lunch hour with the help of several engineer friends, we ran my small turbine of some three inches diameter, up to operating speed of about 60,000 rpm (revolutions per minute) to check it for balance.

For a given power the smaller the diameter of a turbine the higher must be its rpm. Therefore there can be a similar frequency of the number of blades meeting the gas flow per second in a small turbine as there may be in its full size counterpart. In a word the resulting pitch of the turbine scream can also be very similar. In this instance we were running the unit up with a 60 lbs per square inch pressure air line.

Now in the comparative quiet of a huge lunch time hangar.....the noise emitted from this baby turbine was horrendous! So much so that it produced an inrush of startled security police and personnel quite convinced that such an unscheduled run up of the Vampire `goblin' jet engine implied some kind of trouble, if not virtual break in. This small engine, which I designed and built in 1942, is shown here in extracts from my published article on it.

First hovercraft demo.

But of no less interest, it was primarily due to working on this little jet that I first became acquainted with the surface effect of hovercraft. This is the way it happened.

A small gas turbine engine can be very inefficient, in fact it is almost impossible for it to run self sustaining. However it is also known that a pulsing gas flow jet rather than a continuous one can help balance this lack of efficiency. Indeed it is the very same principle as that employed in the German V1 `doodle bomb' jet.

I had figured there was some merit in employing the principle for the small gas turbine, but some experiments were called for. One such arrangement (shown in this section) was investigated in my landlady's kitchen in 1943. This was fed with coal gas from the good lady's cooker (while she was away of course) and was specifically aimed at checking flow rate and the ignition system via a flap valve. The theory being, the heavy weight of an unanchored dish would allow it to pop up and down with each resulting controlled explosion. Subsequent `guestimation' placed the frequency of this to be something in the order of 50 cycles per sec. But for one very brief moment the dish floated and skimmed around the table! Fascinated I had contemplated this phenomenon for a little while before resuming `the more serious work', completely oblivious of the fact that I had just been introduced to my first plenum chamber type hovercraft, which many years later would involve so much of my life.

Sequel panic on the Isle of Wight

What happened to the little turbojet? Well it was completed, but due to temperature restrictions it was unable to run continuously,but still resides in my workshop.Moreover, as a gesture from De Havillands I was presented with a complete general arrangement (G.A.)of the then `secret' Vampire aircraft from which I eventually built a free flight model to take the jet. This model still flies to this day, and it is so realistic that on one occasion, nearly two decades after the turbine session at Hatfield, with smoke pouring from the tail pipe it was seen to make an `emergency crash landing' by several dozen cliff top holiday makers who were totally convinced they had just witnessed a full size aerial drama! I for one have become accustomed to this strange phenomenon of `synchronicity'. By the way, through what persuasive channels did I manage to acquire the Vampire G.A.? It was non other than the young David Kossof, who one day would become a world famous actor!

AEROMODELLER   November, 1948

GAS TURBINE EXPERIMENTS

PART I. By L. G. CRAMP.

ONE seems to hear little or nothing nowadays on the subject of model jet propulsion, and the writer having carried out some research in this field over the last four years, would like to contribute a little toward reviving some interest among modellers.

No doubt among the readers of the AEROMODELLER there are those who will sympathise with the writer, who is often painfully reminded of the fact that the airscrew power driven

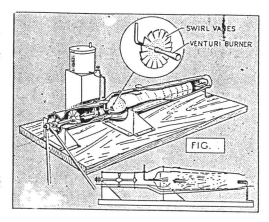

FIG. .

... it was carried out in day-
... a temperature in the region of
By this time the only flame visible was that inside the combustion chamber, this being viewed from the inspection hole provided. It was interesting to note that the flame started several inches down stream from the burner. Now as flame length is one of the many troubles with gas turbine design, it was readily seen that more turbulence was called for in the region of the burner and this was remedied by installing swirl vanes at the entrance to the flame tube. This definitely brought about a shorter flame and after many other experiments it was found that a practical flame length could be obtained by careful fuel adjustment, while still obtaining a high temperature.

After many setbacks it seemed that enough information was gained at least to allow the writer to begin work on the design of the first turbine layout, but before this could be done however it was necessary to have some idea of the working efficiency of the small type turbine and compressor which would be used. A centrifugal compressor was decided upon (as the use of a small axial flow type is impracticable in the writers opinion owing to the high rotational speed required). Again this called for test rigs and for the measurement of the air flow a U type manometer was constructed which was found to give a fairly accurate reading. Now in order to obtain a measure of efficiency for the compressor a rig was required. This was constructed and consisted of a simple train of gears which gave a gear ratio of 4–1 and was powered by an electric motor. The high speed gear had an extended shaft upon which was mounted the compressor under test. A small centrifugal compressor of 4 in. dia., was built of aluminium and assembled to the rig, see Fig. 2, and when started up was found to have a speed of 8,000 r.p.m. The pitot head of the manometer was fixed into position a short distance from the blade tips and when the compressor was running was manoeuvred about a little until the maximum reading was obtained. This position then was the tip exit angle and the approximate angle at which the diffuser blades would have to be fixed and when measured was found to be 44°56, the air velocity being 146 ft. per sec.

Now the theoretical pressure head ▭ that will be produced by a centrifugal compressor is dependent on the tip speed and the blade efficiency and is given by the expression:—

$$Ht. = \frac{U^2}{g}$$
$$Ha = Hty$$
$$V = \sqrt{Ha\ 2g}.$$

where Ht = Theoretical pressure head.   U = Tip speed in feet per sec.   g = gravity 32·2 ft. per sec.   y = efficiency.   V = Absolute air velocity at exit in ft. per sec.

Now as the r.p.m. was 8,000 and the dia. 4 ins. i.e. ·333 ft. the tip speed (U)

$$= \frac{r.p.m. \times dia.\ in\ ft.}{60}$$
$$= \frac{8,000 \times 3·142 \times ·333}{60} = 139·5\ ft.\ per\ sec.$$

And $Ht = \dfrac{139·5^2}{32·2} = 604·3$

Now the velocity measured was 146 ft. per sec. and as $V = \sqrt{Ha\ 2g}$.

$$Ha = \frac{V^2}{2g} = \frac{146 \times 146}{2 \times 32·2} = 330·9$$

$$efficiency = \frac{Ha}{Ht.} = \frac{330·9 \times 100}{604·3} = 54·7\%$$

Now an efficiency of 54% whilst not an exciting figure is a good deal better than some sceptics would have us believe possible with a small compressor of this type, and having received this encouragement the writer built another compressor which gave an efficiency of 56·5%

The next step was the design of a small turbine which would give an equally encouraging efficiency.

(To be continued)

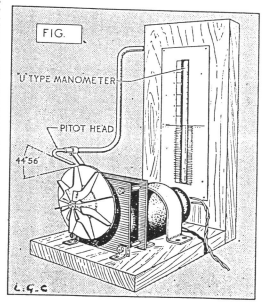

FIG. .

U TYPE MANOMETER

PITOT HEAD

44°56

L.G.C

Extracts from a series of articles published in the UK in 1948.

December, 1948   AEROMODELLER

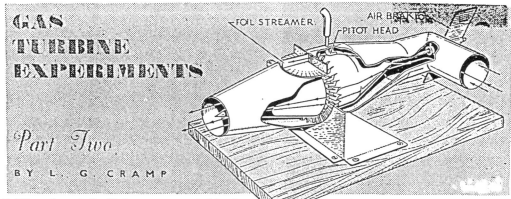

GAS TURBINE EXPERIMENTS

Part Two.

BY L. G. CRAMP

LAST month we dealt with the compressor end of the story and a brief account was given of some experiments carried out together with the results obtained.

And now bearing in mind the difficulty of condensing the information on the gas turbine, without making the subject too hazy, a spot of theory is necessary to allow the reader to analyse the experiments made. One of the first steps toward the design of a turbine, is to have some idea of the working conditions under which it is intended to run, and dealing with these in order we have (a) the maximum working temperature we can allow, together with the pressure ratio obtainable. This will give us some idea of the velocity of the working fluid (Vl) at the turbine nozzles. (b) The maximum rotational speed desirable. Having arrived at this decision we may go one step further and investigate in diagrammatic form, just what happens when the turbine is running. In order to do this we must construct the diagram of velocities, which may be set out to any convenient scale as shown in the illustration Fig. 1. Let AB represent the peripheral speed (U) in magnitude and direction of the turbine wheel, at mean blade diameter. Let CB represent the jet speed (V1) at the turbine stator blades (or nozzles) in magnitude and direction, the angle (∞) is the angle made by the jet with the plane of the wheel. CA then represents the jet (V1R) in magnitude and direction relative to the moving blade of the turbine wheel. Therefore the angle (∞) is the actual entrance angle at which the jet stream enters the turbine, and is therefore the leading edge angle of the blade.

Let angle (B) be that which is made by the exit gases to the plane of ratio of the wheel, then DE represents the exit gases (V2R) in magnitude and direction relative to the moving blade, and DF represents to the same scale (V2) the absolute velocity and direction of the gas stream relative to the turbine casing, due of course to the component EF which is equal to (U) the wheel speed (AB). Now if we redraw the components BC and AC in the position shown by the dotted lines, we shall be able to read off to the same scale the quantity HG which is known as the velocity of whirl (Vw) and one of the factors which will enable us to calculate the thrust on the turbine wheel, as it represents the vectorial difference between V1 and V2, i.e. the absolute inlet and outlet gas velocities at the turbine blade, and since force is proportional to the rate of change of momentum, then the force exerted on the blade in the direction AB by the jet is

$$F = \frac{WHG}{g} \quad = F = \frac{W \, Vw}{g}$$

where W is the weight of gas per second passing over the the blade and g the acceleration force, gravity. This then is the tangential thrust on the turbine blades and it will be observed that this is not as much as the total thrust which is :—

$$F = \frac{W \, FG}{g}$$

due to the fact that there is an axial thrust represented by the component FH, which can be regarded as a loss.

Now the next thing we want to know is a measure of the efficiency of the turbine blade, which is equal to the ratio

$$\frac{\text{the useful energy harnessed by the blades}}{\text{kinetic energy of the nozzle jets.}}$$

Now the kinetic energy of the jet equals :—

$$W \, \frac{V1^2}{2g} \text{ ft. lb. per sec.}$$

and the quantity if useful energy harnessed by the blade equals :—

$$\frac{W \, Vw \, U}{2g} \text{ ft. lb. per sec.}$$

$$\therefore \text{ blade efficiency } YB = \frac{W \, VwU}{g} \quad \frac{2g}{WV_1^2} \quad = \frac{2U \, Vw}{V_1^2}$$

Now if we take ·89 as being the nozzle efficiency (a quite reasonable figure for small nozzles of this type) and multiply it by the blade efficiency, we will then have the turbine stage efficiency i.e. YNYB = Stage Y.

We are now in a position to examine an experiment which was made with a small turbine of 3 ins. mean blade diameter. This was built up of mild steel, the disc (which was of the constant stress type section) was turned up and the periphery slotted to take 30 blades of correct profile.

After being mounted on a true shaft the assembly was balanced and installed into the test rig shown in the illustration. Air delivered from a blower was used to drive the turbine and the delivery pipe and housing were designed so as to prevent a Bernoulli pressure rise; in fact, the cross sectional area from the delivery pipe to the turbine nozzles was slightly decreased so as to increase the airflow velocity.

Now the nozzles in this case were only required to give the airflow the right angle of attack to the turbine blades, and as no expansion would be taking place in them were of a constant cross section and in the test rig shown, these were made up of thin tin plate, and curved through the required angle, which in this case was 42 deg. A small rectangular piece of tin plate was attached to the projecting end of the turbine shaft as an air brake to absorb the power developed.

Velocity diagram and design parameters for the small gas turbine.

January, 1949   AEROMODELLER

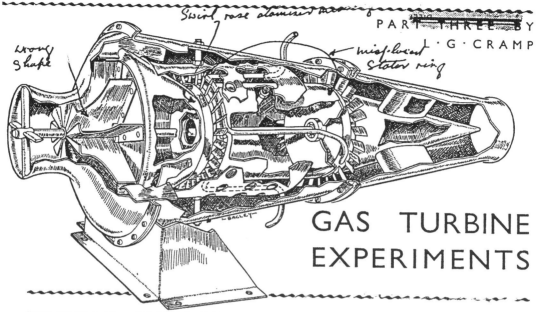

PART THREE BY
L · G · CRAMP

GAS TURBINE EXPERIMENTS

inches and the periphery slotted to take 30 blades. These were made with a true section by first cutting some strips of metal of the required width and pressing while red hot into a channel section to give the right curve, and filing to the correct profile when cold. The strips were then cut into short lengths leaving enough material for the blade roots and tip trimming, and the two corners of the blade roots were removed so that when inserted into the slots, the blade had a true section over its length, leaving two small gaps each side of the blade roots which were stopped up with welding material, the surplus material being removed by filing. The whole job was then put into the lathe and trimmed up, and after balancing and polishing was mounted permanently onto a 3/16 dia., silver steel shaft. Exactly the same procedure was employed for the 28 stator blades.

Now for the fuel system : As vaporised fuel was used pre-heaters were required and these consisted of four small bore tubes, entering the combustion chamber at a point just in front of the nozzle ring and bent with a sharp curve to bring them in line with the flame and after running the length of the chamber, turned through another sharp curve to form a segment of a circle, receiving support from four stops secured to the four tubular bolts, so that with the four fuel pipes in position, a complete circle was formed. Each segment had two small jets drilled into the upstream face, this was found after experiment to prove the best way of mixing the fuel vapour with the incoming air flow. A small spark plug was installed as shown, the earth lead being taken off the casing at a convenient point.

Now as will be seen the incoming air flow offers a small increase in pressure due to the divergent form of the annular entry duct, the primary air flow being bled off from the main stream by means of the small aperture formed by the bell shaped end of the outer flame tube and the conical fairing, and as will be seen, this formed a separate chamber for the primary air and as a rapid divergence takes place at this point, the air flow is slowed up considerably, so that after passing through the swirl vanes and metering rose, mixing is facilitated while the flame is not blown out. This proved quite successful in operation and one of the chief difficulties,

namely combustion, was overcome. After leaving the compressor the main air flow enters the combustion chamber via the holes made in the outside flame tube, the edges of which are projecting in order to catch the incoming air and direct it into the flame centre, while a small percentage of the main air flow is allowed to bypass the flame tube altogether, thereby cooling the products of combustion before entering the nozzles.

Now although the writer anticipated difficulties in the operation of this engine, many were actually encountered, and it was only after months of experiment that any hope of success came into view. The first tests were made of the combustion processes by introducing a large volume of air, delivered from an auxiliary blower, into the compressor intake, the fuel value being slightly open and the spark plug brought into operation, the object of this was to burn a little fuel for preliminary heating, but as the fuel was not sufficiently atomised, this was not possible, but the difficulty was overcome by introducing a small pilot flame. After a preliminary heating the auxiliary volume of air was again brought into operation and the fuel valve manipulated. This time there was no doubt about it, combustion was good, but within a few minutes, the first of the bigger snags arose, namely expansion troubles. This involved a good deal of alteration and trial and error experiments, and when this difficulty had been

January 1949. The last of a series of three articles on the author's small gas turbine showing publisher's sketch.

AEROMODELLER   January, 1949

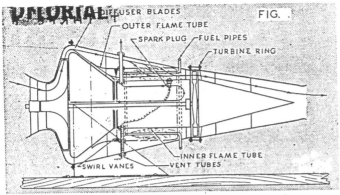

FIG.

DIFFUSER BLADES
OUTER FLAME TUBE
SPARK PLUG   FUEL PIPES
TURBINE RING
INNER FLAME TUBE
SWIRL VANES   VENT TUBES

overcome, it was found that there was excessive burning in the region of the lower part of the combustion chamber and subsequent investigation showed that the cause of this trouble was brought about by incomplete vaporisation of the fuel, therefore there was a collection of fuel in the lower part of the combustion chamber, which burned quite fiercely. Fuel pipes of larger bore and jets of smaller diameter were tried, together with an increase of pressure in the fuel tank, this being some 6 lbs. per sq. inch, and this was found to bring about complete vaporisation and the variation in temperature brought about by a slight adjustment of the fuel valve, was quite remarkable. At this stage a few measurements were carried out which showed only a slight increase in pressure in the combustion chamber, while the r.pm. was in the region of 6,000, so that very little work (if any) was being done by the compressor, so arrangements were made to turn the rotor over with an electric motor at high revs. Now although the pressure was increased by the higher revs combustion was not improved and it was found that the flame length was much too long and instead of terminating before the turbine nozzles, it was found to be half way along the exhaust cone. This was a hopeless state of affairs and the effect it had on the turbine stator blades had to be seen to be believed and it brought home to the writer, with some added emphasis, a remark which was passed to him, in all sincerity, by a gentleman of some authority, when he advised that the stator blades be made of copper!! The unit was stripped down and the blades were cleaned and polished, and the swirl vanes and metering rose redesigned. This also entailed some modification of the fuel ring jets etc., but when the unit was reassembled the results were well worth the trouble. As a point of interest the writer would like to mention that a small turbine of this size does give off a very sweet turbine note, which can be

controlled to a certain extent by manipulation of the fuel valve, this in itself is very encouraging. However, it was discovered that there was a certain critical stage when the engine showed signs of "getting away" and that after a few exhilarating moments, she would slow right up and finally stop until turned over once again by the motor. This indicated that the starting revs were not high enough to allow the compressor to build up sufficient rise in pressure, and another arrangement was made giving a higher speed of rotation. Now at last there was some success, the unit began to accelerate with a high pitched scream, but the temperature was terrific and it would definitely have been unwise to allow this to continue, with the result that only tantalising short bursts could be allowed.

This little unit served its purpose well before giving out and the writer has only one regret, that he was unable to take a measurement of thrust, but many hours of interesting experiment (not to mention a fire or two), were had with it, and at the least has satisfied himself with the knowledge that it is possible for a small gas turbine to function, though admittedly very inefficiently.

At the time of writing this article another design is being prepared, and the writer would be very interested to hear what is being done in this very interesting field.

The following is a table of weights of components parts described in this article.

	ozs.
Compressor impellor	·75
Compressor housing	1·5
Compressor fairing and straightener blades	1·9
Swirl vanes and metering rose	·65
Outer flame tube	2·1
Inner flame tube and stator blades	2·1
Main casing	4·25
Turbine ring	3·
Turbine wheel	3·6
Turbine bullet	2·1
Jet pipe	1·8
Jet orifice	·55
Coolant tubes and nuts	·9
Rotor shaft	1·
Bearings	·85
Fuel pipes	·85
Spark plug	·4
Nuts, bolts etc.	1·25

Total   29·55 = 1·847 lbs.

Even the overall weights have an uncanny close agreement.

Flame-tube

Flame-tube

Gas Turbines 40 years ago??

*Reproduction of a heading to a reader's column
from the winter 1993 issue of the popular Aero
Modelling magazine, **Jet International.***

*The modern realisation(centre)
of the small turbojet engine
compared with the author's
version of 1942 in which even
the flame-tubes are identical.*

*The answer to the readers's question above is **Yes!** This little turbojet measures
just over one foot long, yet it caused near panic and consternation at the De
Havilland Aircraft Co. factory at Hatfield, Hertfordshire, UK in **1943.***

*Author's son Gary only eight years
at the time..*

*The very first free flight model
of the De H. Vampire, designed
and built by the author in 1947,
from the makers original plans,
which caused bewildered
consternation among onlookers
at a UK coastal holiday resort in
1983. Note the prototype **square
cut fins**. Compare with recent
model below. The author's work
on this model has never before
been publicised.*

Pete Marsden, the first pilot to fly a
JPX powered model in the UK, has
recently built this own-design DH
Vampire as a testbed for a future more
'scale' version. The model is 76"
wingspan, weighs only about 13Lbs and
flies like a 'trainer'. Watch out Pete's
article on the new scale version in a
future RCJI, together with some clever
weight saving construction techniques.

*Time lapse and compounded synchronicity. Why of all model aircraft in which to put a
new breakthrough miniature turbojet should Pete Marsden have chosen the De H.
Vampire and the prototype version at that?*

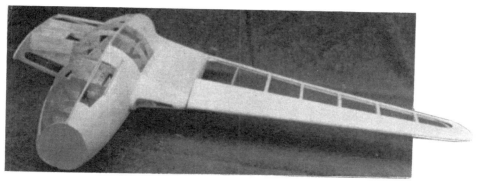

Author's partly finished model of the prototype Vampire showing balsa planked structure. (fibreglass wasn't in vogue in those far off heady days!)

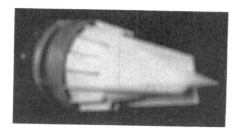

The original power unit for the Vampire, probably the first functional centrifugal type ducted fan on record. Commercial ducted fans didn't arrive until over three decades later.

Showing balsa-ply frames, skin planking and original Mills diesel engine, which still resides in the author's workshop!

*Exploded view of the ducted fan
unit minus tail pipe, looking aft.*

*Forward exploded view.
Central diffuser houses
centrifugal fan.*

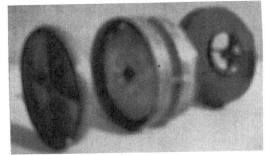

*Exploded view of balsa constructed
ducted fan diffuser.*

Front view of centrifugal/axial fans.

Rear view of fan.

*The author's centrifugal/coaxial ducted fan unit weighed under one
pound and turned at over fifteen thousand RPM.*

An anecdote

In those WW11 days an aircraft establishment such as De Havillands was a fairly modern set up complete with special police guards at each entrance with rising barriers and white painted pill boxes. You simply couldn't get in or out of the place without a departmental pass. Indeed there was something very restrictive about it, much like a PoW camp, or dare one say, similar to present establishments like Area 51.

When I was first introduced to David Kossof I very soon discerned he was regarded by most people. who had a need to know, as the best `Mr. Fix it'. It was amazing just how much he could `arrange' for a packet of cigarettes, again just like his opposite number to be found in most PoW camps. It so happened one of his specialties was bogus departmental passes. These were specially printed chits requiring only a sanctioning signature from the appropriate departmental chief. Not only was David a first class actor (he used to appear at the Unity Theatre) he was a first class artist and his flourishing, very authentic looking, totally unreadable, doctors type signature, never, never failed to convince the police guards. I don't know how one would describe this `art', but no one could say it was forgery in the strictest sense.

Granted the security system at De Havillands was a very necessary one, nevertheless sometimes it could be rather daunting when for example, having earned a weekend break, one ended up waiting precious hours in an unlit, unheated wartime railway station, simply because you missed the connection due to an overloaded firm's bus letting you down. So a few minutes early departmental pass was a godsend, the snag being you were allowed only one a month. A restriction which rather unfairly did not apply to those bosses, who like several of us, had extremely long cross country journeys to make. I don't know from where or how David obtained his private stack of pass slips, one didn't ask, the proverbial packet of cigarettes did the rest.

It was amazing to watch the superior air with which the police guard dutifully scrutinized these `official' documents, as if this very act of authority had just made his day. Thankfully I did not have to do this very often and when on occasion I did, I felt as guilty as sin and hoped like blazes I didn't show it. But the thought of those cold dark often midnight hours on a bleak deserted railway platform could work wonders for one's self confidence.

The last time when David came to my rescue in this way was on an occasion when perchance we both had the opportunity to catch an early train. There had been no time to spare, our weekend bags were packed, and we were wearing our weekend travelling suits, but David had run out of `spare' pass slips! He had glanced at his watch muttering "We might just make it", ushered me out of the office and we were on our way. Utterly confused I was at a loss to understand why we were heading in the direction of the furthest exit. But it soon became clear when he assured me the police on duty at that particular point were casual staff. To my frantic "But how are we getting out without passes?" he merely assured me it would be all right. By now I was all for going back, but David had persisted, instructing me to act normally and not be surprised by anything he did or said. Very soon we were approaching the awesome pill box. I saw the guard lay aside his newspaper and pipe, remove his reading glasses, stand up straightening his tunic prior to intercepting us when David whispered "Now leave it all to me". Instantly his normal slightly cockney yiddish accent vanished and in a prominent extremely imposing cultured voice said "But my dear Leonard, as I was saying to Sir Geoffrey (De Havilland) only this afternoon......." Whereupon that police sergeant touched his cap respectfully, said good afternoon gentlemen, raised the barrier and let us through.....! That may seem to have little to do with aeroplanes, UFOs or the paranormal, but it was an extremely impressive spontaneous demonstration of real acting, and I have never forgotten it.

Balsa wood, balsa wood galore

When I was still a commercial artist my love for model aircraft had never waned, but as most of my modest income went towards the family budget there was very little money for me to purchase very expensive balsa wood for construction. So having joined De Havillands I had been dumbfounded to learn that the huge balsa wood waste from the aircraft production lines was sold at scrap value prices to the employees. In different grades, there were assorted sized blocks and sheets, and each purchaser was allowed a total of 5 cubic feet! Five cubic feet of random sized wood can represent a fair sized `parcel' which was the envy of most local stockists. Needless to say, the aeromodelling club at De Havillands was *very* popular. Somehow I managed to convince a somewhat doubting landlady that this `stuff' was extremely valuable and I eventually managed to transport it home to Norfolk. For the next few years I was not found wanting for balsa wood, indeed it is appropriate that the model DH Vampire shown in this chapter, was built from it,

Of course the apprentices at De Havillands had their share and among them were two youngsters whose names in the future would become very famous, they were non other than Desmond Norman and John Britten of Britten Norman aircraft fame, who were to give the world the `Islander' one of the most successful aircraft still in use today.

It is interesting to note that while they were engaged in their more traditional apprenticeship at Hatfield, I was continuing my particular type. One thing we had in common though, without prior intent our paths would one day converge on the Isle of Wight. It will be seen later that one day I would be employed by Desmond Norman in his Fire Cracker design team and as a side line design and build scale flying models for his brother's toy factory in Scotland, not least as a test pilot Desmond accepts the reality of UFOs, while Piece for a Jigsaw has a place of importance on his book shelf.....synchronicity or not, it's a small world isn't it!

1956 - 1985 From small beginnings the Cold Turbine ram-jet.
ATF 29 years.

The work on the small turbojet was directly responsible for the Cold Turbine Ram-Jet in 1956. Although the more recent look alikes are practically identical in basic concept, there is no known reason to doubt that these latter versions are completely original. The following short article was originally specifically intended for this present book. However recent disclosures have warranted that it should be included in my last book The Cosmic Matrix. The author acknowledges this policy together with the hope that it will cause no confusion to readers.

`From little acorns grow'

In 1948 I joined the design office at D.Napier & Son Ltd., at Acton in London,UK. As is well known their specialty was engines, ranging from large `Deltic' compound diesel engine - used world wide in diesel electric locomotives - to smaller more powerful turboprop and jet engines.

1956 found me attached to the advanced projects department (APD) where we often had to graphically complete certain areas of the designs which required rapid attention in order to meet a deadline date - usually tomorrow - for important meetings with various dignitaries from such as the Ministry of Defence. One design designated the HTU 1 (Hot Turbine Unit) involved a rather costly, convoluted and complicated solution for cooling turbine blades by a process of forming them hollow in order to allow cooling air to be pumped through them, a process which involved conveying the air through a very clever series of labyrinth seals. The idea was sound enough but extremely tedious to graphically portray - we were not blessed with graphic displays and computers in those days.

In view of the short time available I was not overpleased with this somewhat daunting task, nevertheless it did have the effect of spelling out to me the extreme lengths designers were prepared to go to in order to increase gas turbine working fluid temperatures (WFT) by only one or two degrees celsius.

On my journey home I was pondering the problem in terms of some of those I had met years ago. I found myself contemplating some of the technical compromises gas turbine engineers had to make, and if the turbine power output could be *considerably* increased by only raising the WFT by a degree or so, just imagine what output could be achieved from the same engine if the WFT could be significantly raised. Indeed suppose the flame temperature need not be diluted *at all?* My mind boggled as I thought `if only they could do *that'*.

Of course by this time I was sufficiently skilled in the art to be able to take the thought further with a few basic sums and some rough sketches. The conclusions were disturbing, for I couldn't fault the idea. But I was equally impatient because it was so obviously simple, engineers all over the world would have tried it, for if my hypothesis was correct it implied jet propulsion turbines could in fact be formed from light inexpensive alloys, for with such a design they could *run cold!* Sometime I would curb my pride and ignorance and try to find out. That time turned out to be sooner than I thought.

The following day I had to direct all my attention to getting the job on the HTU 1 finished in order to meet the deadline. But this was not easy, for working on the complex system only served to rekindle the thought `why...why not?'

During the morning, Mac the technical writer working on the job with me, came over to discuss matters several times and I found myself impatiently wishing to put the question to him for he had a doctorate in thermodynamics and was certain to be able to quench my disquiet. As the morning wore on I finally yielded and apologizing for the digression, showed him my rough notes with the question, `as there *must* be a reason, what was it?' This was followed by a somewhat disturbing silence and a brief moment or two of hesitance, then finally he said "I don't know!"

Several days later we had finished our allotted job on the HTU 1 in time, our visitors had returned to the city highly pleased, our boss was relieved and so were we.

Not one but three Thermodynamicists

Around three o'clock in the afternoon Mac came in to see me, rather excited and slightly flushed I thought, and said "With regard to your question the other day, per chance this lunch time I sat down at a table occupied by not one, but three APD thermodynamicists" He paused and then said "I hope you don't mind, but I put your question to them". Sure I didn't mind, if anything I was glad to have a chap of his technical calibre at least willing to be told, as much as myself. I waited patiently, then in response to my questioning glance he said "that's just the point, it's now three o'clock in the afternoon and I have only just left them....*arguing among themselves* about it!"

For several nights I worked long into the small hours as I have always done, building a fully sectioned table top display model engine complete with rotatable rotor and pretty stand, I had a hunch I was going to hear more...and I did.

Breakthrough & Dr.Morley

Within a day I was called to the `higher sanctum' in the firm's technical division at the request of the then chief advanced projects scientist Dr. Morley, on assignment from the Farnborough Aerospace Establishment. My own immediate superior was enjoying himself, bathing a little in the `reflected glory' as he gave me the news, `just imagine one of his chaps with a major breakthrough potential'.

Dr.Morley's secretary announced my arrival and I was cordially greeted by him sitting behind an impressive desk in an equally impressive room, with walls hung with a variety of pictures appropriate to the aerospace industry. He rose to meet me, shook hands and proffering a chair came round and sat on the edge of his desk lighting a pipe in the process. Then after formally enquiring about my background the questions began. How long had I worked in the projects department and what motivated the idea etc.? Dr.Morley seemed curious to know as much about the cold turbine'' inventor as the concept itself.

Eventually I handed the box containing the display model to him and one of the fondest memories I cherish is the broad grin of appreciation on his face when he raised the lid, and I was immediately glad for the long night hours I had spent building that model.

He duly thanked me for it, saying he would be delighted to use it at future meetings as he had already placed the idea in the hands of several jet propulsion specialists in the APD. However I was entirely taken aback by Dr.Morley's next question when he asked if I would be willing to join his team in the APD as he considered I was something of an `ideas man'. I cordially thanked him, but had to tell him my family and I were considering moving to the Isle of Wight, and in any case, as higher mathematics was not my forte I might prove to be inadequate for such a position. He said that my intended move was a pity, but as far as my rusty maths was concerned he would be glad to make arrangements for me to polish up at Napier's technical college a couple of times a week.

`The wood for the trees'

During the period which had elapsed since my first conversation with Mac, I had little time to devote to curiosity, yet I was puzzled somewhat insofar as surely my proposal was more the province of specifically trained engineers? Leaving Dr.Morley's office that day something he said made me hesitate with that unanswered thought still lingering in my mind. I shall never forget the image of that kindly man sitting there with arms folded and pipe in hand he too hesitated for a moment then smilingly said "It's simply a matter of all too often we can't see the wood for the trees!"

A new future had been offered me, a self taught aerospace enthusiast, and although it was a tough decision to take, my family and I did move to the Isle of Wight where the pull of a job working on Britain's first space rocket was too much to resist. I would have been honoured to work for Dr.Morley and share with him some of the magical things it has since been my good fortune to find.

It is my sincere wish that this short account may in some measure help to reassure other, equally not so well trained, but insight gifted youngsters not to be afraid to ask "Why?", as this retort from an honest man may convey. Since those days I have seen the truth in that stance upheld many times. However from time to time we all suffer from `wood for the trees' myopia. I certainly know I have, and the truth in that metaphor has not weighed lightly in the decision to present results of my work on gravity and energy in The Cosmic Matrix.

Liquidation and missing documents

What happened about the `cold turbine' jet engine? The Ministry of Defence clapped a top secret notice on it, I met with officials several times. Then the project was taken over by a subsidiary firm, English Electric at Luton Airport, Bedfordshire, UK for wind tunnel tests, where it was eventually shelved, `as the faculty had no access to a hypersonic wind tunnel', which was then unavailable in the UK. Eventually D.Napier & Son went into liquidation, therefore in hindsight I did the right thing in resigning my position and moving to the Isle of Wight. Subsequent events deemed it to be so.

I never heard another word about my cold turbine design although Mac did manage to rescue a copy of the patent application (shown in this section) together with a copy of the original design study from the APD. Some years later this was found to be missing from my workshop suspiciously close to a visit from representatives of a large overseas concern who were at that time showing interest in another of my projects.

A cold turbine ejector engine

A new form of gas turbine engine, in which the turbine is relieved of high gas temperatures, is proposed in this article. Protective rights on this, the author's own idea, have been filed by a major aero engine manufacturer

IN the aeronautical world it is natural for the word turbine to be associated with thoughts of intense heat, and we accept the fact that it is chiefly the turbine and the stator ring in a turbojet engine which takes the full blast of the hot expanding gases.

The flame temperature of a modern gas turbine can be in the order of 2 000 degrees C, and the gases have to be 'diluted' and cooled down to a temperature of about 850 degrees C before they pass through the stator and turbine. This is known as the 'working fluid temperature' (w f t) and upon its magnitude largely depends the efficiency of the cycle.

The w f t is usually governed by the temperature at which the turbine can operate, and when it is borne in mind that the turbine in a modern turbine engine runs red hot—suffering extremes of centrifugal and bending loads in the process—some of the problems which have beset the metallurgist can be appreciated. A turbine blade which, at rest, weighs only a few grammes, can experience a disrupting stress of something like 15 kg/mm² (10 ton/in²) due to centrifugal force alone.

Naturally, under these conditions the blade is inclined to elongate which, if unchecked, can—and does—lead to disastrous consequences. Largely due to the enormous strides made in metallurgy, which has furnished gas turbine designers with better materials, the permissible w f t has been pushed higher and higher, and for each small increment of temperature, so has the power output of engines soared. If the inlet temperature to the turbine can be raised, not only will the thermal efficiency be increased, but the quantity of air required for cooling the combustion products will be reduced progressively, so that a smaller compressor and turbine can be used for the same net power output (see Figure 1). It is to this end that designers are constantly striving.

One of the most promising solutions to the problem currently receiving attention is to form hollow turbine blades into which is fed cooling air. This, unfortunately, can only be obtained by further complexity, which, of course, incurs weight, manufacturer's costs and maintenance penalties.

In this arrangement air is 'bled' off the main compressor staging and is induced to flow through various intricate labyrinths and passages to the main compressor turbine rotor shaft, and then to the turbine disk itself. Further passages in the disk convey the air through the roots of the hollow turbine blades, where—now considerably heated—it passes through small outlet holes situated in the blade trailing edge, as in Figure 2.

Research and development work on these lines is going on throughout the world, while the metallurgist continues to find stronger and greater heat-resisting alloys.

From all this it will be appreciated that should it be at all possible to extend the w f t any higher, even greater power could be attained.

L G Cramp

The following brief report on the writer's attempts to find a solution to the problem may suggest a possible way of achieving maximum w f t, though it is pointed out that the principle outlined here requires further examination. Early enquiry into the cycle's functional capabilities has indicated that it would work, though it might have limitations which need not be prohibitive.

It is now well known that in the conventional gas turbine cycle the pressure energy contained in the hot gases is converted into kinetic energy in the form of an expanding jet of high velocity gas, which impinges on the turbine blades giving up some of its energy to the rotor system and releasing the remainder through the propelling nozzle for jet reaction.

The 'ejector' engine, as the writer's engine is called, is somewhat different in operation, deriving its name from the ejector principle. As with the ramjet it offers the advantage that gas dilution is unnecessary, inasmuch as the products of combustion never come into contact with the turbine at all. Instead, provision is made for the rapidly expanding high velocity jet to move a large volume of cold air, in which is placed an air turbine.

The principle is basically simple and automatically suggests the engine's component layout, one of which is shown diagrammatically in Figure 3c.

The original conception (Figure 3b) may serve to convey the natural development of layout. In this configuration we have a conventional turbine-compressor assembly as in Figure 3a, but with the exception that, at the nozzle box, where the hot gases enter the stator blade ring, there is placed a duct leading to atmosphere. Although it will be appreciated that this is an unworkable arrangement, it may serve to convey the general idea in its broad sense.

It will be apparent that the high velocity gas jet will induce an inflow of cold air, giving up some of its temperature to the air in the process. Therefore we could expect to be able to increase the w f t. Indeed, this would be necessary, for the hot gases would also have given up some energy. In addition, the turbine would still be exposed to heat, so there would be no obvious gain in this direction.

But the cold air entering the duct has considerable energy, energy which was given to it by the expanding gases, which would normally drive the turbine direct. Theoretically, therefore—allowing for losses—the volume of cold air should have almost enough energy to drive the turbine compressor set. And in any case we can now increase the w f t to a maximum.

Re-arrangement of the components

Figure —This graph indicates the improved overall efficiency of the cycle which occurs when the temperature of the working fluid is increased

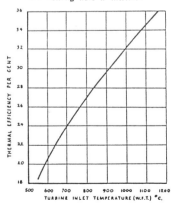

THERMAL EFFICIENCY PER CENT

TURBINE INLET TEMPERATURE (W.F.T.) °C.

By 1960 the author had moved to the Isle of Wight and due to the shelving of the cold turbine project I decided to publish this article on it.

gives the layout shown in Figure 3c. From this it will be seen that the turbine and compressor are integral, so arranged as to form two-tier blading on a common rotor disk. The tiers are separated by a ring which separates the blades and also forms the continuation of the turbine-compressor intake ducting, Figure 4.

Figure 3a shows, for comparison, a conventional jet propulsion gas turbine layout of similar compressor and turbine staging, while the structural and general advantages, which chiefly affect size and weight, are as follows :—

(1) The overall length of the engine is greatly reduced.

(2) The costly and heavy rotor shaft is eliminated.

(3) The turbine rotor, being integral with the last stage of the compressor assembly, represents a further saving in manufacturing and assembly costs.

(4) Flame length, which seriously influences overall size in the conventional turbine engine, is of little or no importance.

Although these represent the more ious advantages, closer consideration will reveal others of not so dramatic a nature, but which are enhancing, nevertheless.

The starting procedure of the ejector type engine would be similar to present-day turbojets. The rotor is motored up to operating speed, drawing in air which suffers a rise in pressure as it passes through the compressor stages. Passing into the diffuser accompanied by further pressure increase it is delivered into the

Figure —This illustration shows the arrangement of the two-tier blading. The guide vanes and stator rings are omitted for the sake of clarity

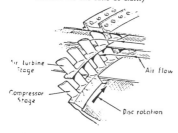

combustion chamber as the primary airflow.

Fuel is then burned in the normal way and the ensuing products of combustion are expanded at high velocity through a convergent type nozzle into the mixing zone.

The considerably large volume of air thus expelled from the rear ' mixing ' chamber causes a pressure drop in the air turbine annulus, which in turn induces a high velocity flow of cold air across the turbine. The energy imparted by this secondary airflow to the turbine is absorbed by further work done in the compressor and the cycle is then completed.

Effectively then, the energy in the products of combustion are absorbed by the turbine as in normal practice, except that, part of this energy is employed to move a protective belt of cold air which, in turn, operates the turbine. As already stated, this represents, to some extent, a drop in the efficiency of the cycle, but it must be remembered, on the other hand, that the w f t can now be maintained at a maximum, requiring no further dilution.

A hypothetical engine has been designed to meet these conditions, and all but one stage of the axial compressor has been eliminated. This, in effect, reduces the moving parts of the engine to little more than one two-tier rotor, the inner ring comprising the compressor stage, and the outer, the turbine. A similar arrangement to this was employed on several ducted-fan augmented jet engines.

In Figure 5, normal stator rings are included for both turbine and compressor, and it was anticipated that variable inlet guide vanes would be necessary.

The writer—having successfully carried out experiments with small centrifugal type fuel injectors—is of the opinion that this type of pump would suit the basic simplicity of the ejector engine ; the remainder of the entire unit could be fabricated.

Should it be proven that the engine cycle would not sustain itself under static conditions, that is, it would require some ram effect from forward motion, then perhaps the addition of a small liquid fuel rocket might prove an attractive alternative, as illustrated in Figure 5.

Such an arrangement should prove to be quite a space-saver in mixed unit type aircraft. In this case, the cycle of operations would be for the rocket to be fired and the rotor ' motored ' up to operating speed simultaneously ; the resulting increased mass flow through the engine being employed not only to accelerate the vehicle to its critical

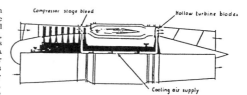

Figure —Cooling air is bled off the compressor and led to hollow turbine blades

Figure —The three drawings below indicate the natural development of the ejector engine

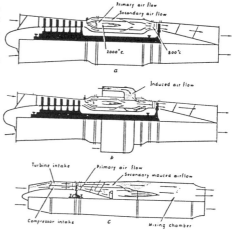

ram speed, but to offer further increased thrust for take-off.

At the time of writing there is news of a new American turbo-ramjet propulsion unit, and although this title would seem to convey a similar idea to the ejector engine it is possible that it differs in its basic concept.

• Results of a preliminary design study by D Napier & Son Limited revealed that the economic operational margin might be restricted and, in fact, it was shown that the simple cycle would not be self sustaining under static conditions. However, the study also revealed that the ejector type engine would yield a thrust co-efficient and a specific fuel consumption comparable with a ramjet at transonic speeds. Protective rights for the ejector engine have been filed by D Napier & Son Limited.

Figure — A cut-away drawing of a rocket-start ejector engine

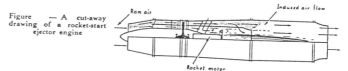

ᴘᴬTENT SPECIFICATION

DRAWINGS ATTACHED

Inventor: LEONARD GEORGE CRAMP

827,663

Date of filing Complete Specification Nov. 1, 1957.

Application Date Sept. 3, 1956.

Complete Specification Published Feb. 10, 1960.

No. 26922/56.

PATENTS
DEPARTMENT

D. NAPIER & SON
LIMITED
W.3.

Index at acceptance: —**Class** 110(3), G(1A : 19), J1A, J2(A1A : A1C : B4 : BX).
International Classification: —F02c, k.

COMPLETE SPECIFICATION

Internal Combustion Engines

We, D. NAPIER & SON LIMITED, a company registered under the Laws of Great Britain, of 211, Acton Vale, London, W.3, do hereby declare the invention for which we pray that
5 a patent may be granted to us, and the method by which it is to be performed, to be particularly described in and by the following statement:—

This invention relates to internal combus-
10 tion engines and is primarily applicable to aircraft propulsion units.

An internal combustion engine of the continuous combustion type according to the invention comprises an air compressor
15 arranged to deliver air to a combustion chamber, and means for delivering fuel to be burnt with the air in the combustion chamber, the hot combustion products being allowed to expand and discharged at high velocity
20 through an exhaust aperture, the compressor being driven solely by an air turbine arranged in a relatively cool air stream created or assisted by the high velocity gas flow discharged from the combustion chamber.

25 Thus an important advantage of the present invention over a conventional gas turbine engine where the air compressor is driven by a gas turbine arranged in the hot exhaust stream, is that the turbine is subject only to
30 the relatively low temperatures of the air stream.

According to a preferred feature of the invention the air turbine lies in an air duct which communicates at its downstream end
35 with a common exhaust duct into which the hot high velocity gas flow from the exhaust aperture also passes, the junction between the two ducts being in the form of an ejector such that the high velocity gas flow
40 induces or assists the flow of air through the air duct.

Where the engine is mounted on an aircraft to act as a propulsion unit the air flow through the air duct will of course be ampli-

fied by the "ram" effect of increased air 45 pressure arising from the forward velocity. In fact it is thought that some "ram" effect will be essential for the cycle of the engine to be self sustaining.

According to another preferred feature of 50 the invention the compressor and turbine are both of the axial flow type and are mounted on a single rotor assembly in the form of radially spaced tiers or rings of blades. This simplifies the construction of 55 the rotor assembly and the air duct is then conventionally arranged as an annular passage surrounding the combustion chamber and communicating with a common exhaust duct at the downstream end of the chamber. 60

Means may also be provided for supplying fuel and burning it in the gas flow downstream of the exhaust aperture.

The invention may be performed in various different ways but one specific embodiment 65 will now be described by way of example as applied to an aircraft propulsion unit of the jet propulsion type, illustrating in section in the accompanying drawing (which is partly broken away and shortened for convenience). 70

In this example the engine comprises an outer elongated generally cylindrical casing 10 which is tapered slightly inwards towards its upstream end 11 (which constitutes an air intake) and within which is mounted an 75 inner cylindrical casing 12 extending lengthwise over a limited part of the total length of the outer casing and spaced appreciably from its two ends. The inner casing is supported from the outer casing by three series 80 of stream-lined spats or struts 13, 14, 15 to provide an outer annular air duct 16, surrounding the inner casing.

Adjacent the upstream end of the inner casing 12 a central bearing assembly 17 en- 85 closed within a bullet-shaped fairing 18 is supported on inner extensions of the radial spats 13, 14 and in this bearing assembly

[*Price 3s. 6d.*]

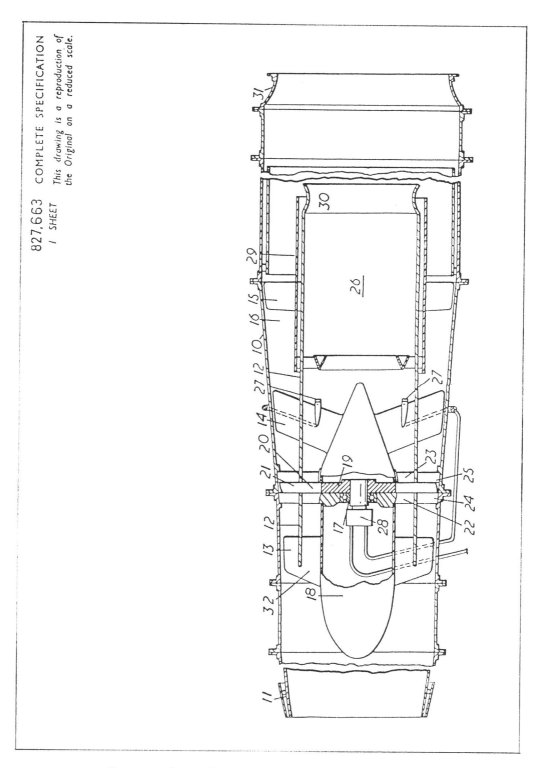

827,663   COMPLETE SPECIFICATION
1 SHEET   This drawing is a reproduction of the Original on a reduced scale.

Patent specification drawing for the cold turbine ram-jet engine.

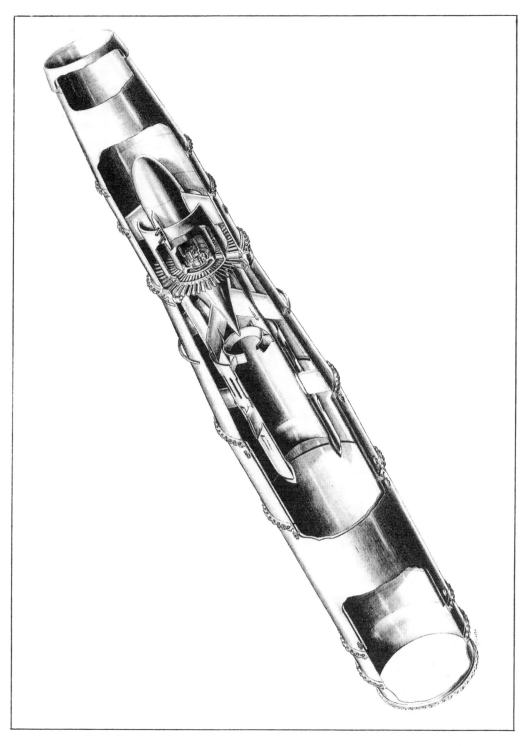

*Author's orthographic scale projection of the cold turbine ram-jet drawn for
D.Napier in 1955. The start rocket motor has been omitted for clarity.*

Continued from page 16
Extreme stresses may also be present owing to the speed and weight of the vehicle. For craft travelling at up to Mach 5, titanium or advanced aluminium alloys are likely to be used, probably shaped using superplastic forming techniques.

At greater speeds, carbon-carbon or ceramic matrix composites, and double-skinned metal matrix composites, are among the materials proposed. Again, small components such as Space Shuttle wing leading edges (manufactured from carbon-carbon) are already in service. The technological challenge will lie in adapting laboratory processes for large-scale manufacture.

British Aerospace's Hotol concept, first presented at the 1984 Farnborough Air Show, is very similar to some of the US proposals. At the time of writing, BAe is awaiting around £2million-worth of government funding for a two-year proof-of-concept study into the viability of the spaceplane. If the money is forthcoming, BAe and Rolls-Royce hope to bring Hotol to the point at which it is seen as a programme worth full-scale funding. If the project does not then appeal to the European Space Agency, it may be offered as the basis for partnership in a US programme—particularly if Rolls-Royce's liquid-air-cycle engine concept proves workable.

Darpa's current timescale for a hypersonic transport programme envisages engine/airframe evaluation over the two years beginning in January 1986, construction of a flight-test vehicle taking place in 1988-1991 and flight-testing in 1991-1995. Around $3,000 million will be needed to reach the flight-test stage.

An airturboramjet (ATR) seen undergoing static test at Aerojet's Sacramento, California, facility. The engine combines turbine, rocket, and ramjet technology, operating like a turbojet subsonically and like a fan-boosted ramjet at supersonic and hypersonic speeds.

The ATR's atmospheric cycle is similar to that of a conventional turbojet. The engine's core contains a compressor to increase the ramjet pressure ratio. This is driven by a turbine, which, instead of being open to the airflow through the engine, is enclosed in the powerhead housing, where it can be driven by a high-pressure fuel-rich hot gas from a separate combustion chamber. Fuel is partially burnt in this chamber, then passed to the nozzle chamber for final combustion. Alternatively, cryogenic fuels can be used to provide active cooling of flight surfaces, becoming vaporised in the process, the vapour being used to drive the turbine before being burnt in the nozzle.

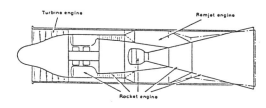

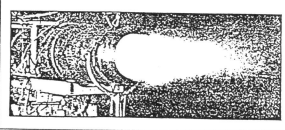

FLIGHT INTERNATIONAL, 7 December 1985

The close similarity to the author's original cold turbine aerospace engine concept of 1960 and that now being investigated in the UK and USA as an 'exciting new idea of breakthrough proportions' is apparent from this article reproduced from Flight International 1985 and more recently in 1989 in th Daily Telegraph.

16   THE DAILY TELEGRAPH, MONDAY, APRIL 24, 1989

SCIENCE & TECHNOLOGY

Inventor Alan Bond hopes a secret new engine will give extra thrust to his precious space plane project

Could Satan help Hotol take off?

SPACE TRAVEL

Adrian Berry

ALAN BOND, frustrated inventor of the space plane Hotol, is not defeated yet. While the project has collapsed because of the Government's refusal to fund it, he has now designed a new engine for the craft. Unlike the earlier design, whose patent he sold to Rolls-Royce, the new one remains his own property, to sell to whom he chooses.

"It's called Satan," he said. "It's significantly more efficient than the original Hotol engine, but I'm not revealing the details to anyone unless I can sell it. I can't even reveal what Satan stands for. There's nothing sinister about the name, but it is an acronym that might provide clues as to how it works, but it shall not patent it, since that would risk it being classified. I believe the Government has no legal power to stop me from selling it, even to a foreign government, if that should be necessary."

British Aerospace's design for Hotol are revealed on this page. If Satan gets off the ground, Hotol (short for Horizontal Take-off and Landing) will carry cargo into space at less than a fifth of the cost of the US space shuttle. And for a third of its total construction cost, it could be upgraded to carry 20 to 30 astronauts, three times more than the shuttle can carry.

Satan is similar in size to the earlier Hotol engine, so there is no question of having to redesign the overall 275-ton vehicle. Omitting secret details, it will save weight by flying with a "ramjet" air-breathing engine that will take its fuel from the atmosphere itself.

It will take off horizontally like an aircraft, boosted by an engine component fuelled by liquid hydrogen. It will then climb with its ramjet to a height of some 18 miles (six miles higher than the cruising altitude of Concorde) where it will have reached six times the speed of sound. There, where the atmosphere becomes too thin to provide any more fuel, its on-board liquid oxygen engine will ignite, accelerating the craft to 25,000 mph, the speed necessary to reach orbit.

It is far less wasteful than Europe's Ariane rockets, since, unlike them, it is wholly reusable. It would land on an airport runway without jettisoning any of its parts, just like an airliner.

Satan gives it a new dimension. "We haven't yet been briefed about this new engine," said Dr Robert Parkinson, manager of future launch systems at British Aerospace, "but it sounds as if Alan has something good up his sleeve."

British Aerospace, which still hopes to manufacture and market Hotol, has 60 people working on it, and they have just produced the new detailed blueprint of the whole craft. "An improved engine design will certainly make Hotol more viable," a spokesman said.

Bond, now 45, has spent all his life designing rockets, from what were little more than child's toys to the engines of starships. None of them has yet flown. While still a schoolboy, he saved up his 7s 6d a week pocket money for several months to build Poltergeist, a 4 ft 6 in projectile that he planned to launch from a 19 ft tower on the Derbyshire moors. It was going to fly to a height of 10 miles, to beat the American amateur record of the time.

Forbidden to fly it by the Air Ministry because of the Explosives Act of 1875, and threatened with a fine of £100 for every day the rocket remained in his possession, he turned his mind to spaceships of the distant future.

In a team formed by the British Interplanetary Society, he helped to design Daedalus, an unmanned starship crewed by intelligent robots that, during the 21st century, may fly to Barnard's Star, six light years from Earth, in a one-way trip that would last 50 years cruising at 84,000,000 mph, 12 per cent of the speed of light.

He also published in a learned journal a blueprint for Rair, short for Ramjet Augmented Interstellar Rocket, a far-futuristic version of Hotol that would work by nuclear fusion. It is an even more spectacular idea than Daedalus. With thrust increased by injection of the elements cadmium and thorium into its reactors, it would suck in the hydrogen fuel of interstellar space for its starship engines at space for its starship engines at cubic miles around the ship, giving it an estimated speed of 70 per cent of the speed of light.

Ilis Satan engine is his last attempt to build a successful rocket. "If I can't sell this one," he said, "I shall really give up. It's my last throw." None of his friends believes this.

THE CHANGING SHAPE OF HOTOL

British Aerospace original concept

Carbon-poly-ether-ketone sandwich (Carbon PEEK)

Liquid oxygen tank

All moving fin

Carbon silicon carbide nosecone

Gravity operated steerable nosegear

Jettisonable titanium foreplanes

Liquid hydrogen (LH₂) forward tanks

Baffles of carbon PEEK

Thermal protection shield panel

Liquid Hydrogen aft tank

Tank divider

Variable geometry intakes feed atmospheric air to fuel engines

Payload bay

Payload bay door

Eight heat exchangers

Engines switch to liquid oxygen fuel at high altitude

Wing liquid oxygen tank

Size comparison of Hotol to Shuttle

Graphic: ROY CASTLE (with acknowledgments to FLIGHT INTERNATIONAL)

Daily Telegraph 1989
The proposed Hotol space craft which employs the `new ram jet' engine.
Note the bracketed ram jet description above.

1958 - 1975 The beginning of a hang glider. The Aeroski one of the first proposals on record for the flexible wing as a commercially available hang glider.
ATF 17 years.

The first friendly gift

The following narrative concerning the Aeroski really had its origins way back in the 1930s when as a young lad I acquired my first model sailing yacht. In those early times my fascination for flying had extended to include sailing, because somehow I recognized an empathy between the two.

Being unable to afford such luxuries I compromised by sometimes visiting Wanstead Flats in Ilford at weekends, to watch model sailing events on the beautiful lakes there. I remember the feeling of awe, if not downright envy I suffered watching the likes of retired sea Captains adjusting their beautiful large models of six feet or so. But now that I was older I could ask people to borrow books from the library for me on the subject of yacht building. Very soon I was able to loft from sections etc.,and I scrounged empty tea chests from the local dock yard and the like for the plywood to build with.

It so happened my father was a gifted engineer and mechanic, who unable to afford tools like planes, spoke shaves and drills etc.,had made his own, examples of which I still have to this day. Also he did freelance night work in his shed at the bottom of our yard and it was there I cut my teeth on model making.

In summer months, not being able to sail models of our own, my friends and I used to swim out to an island in the lake, and it was during one of these escapades one of the boys found the partly submerged wreck of a large model motor yacht. As this was of little use to him, despite its poor condition and to the disapproval of everyone, he retrieved the wreck and kindly somehow managed to bring it home for me.

Having discussed this `project' with my father I set about redesigning. First the solid prow was reduced in length by three inches and reshaped to conform to the profile of a racing yacht. Next the damaged and partially parallel mid section of the boat was removed and the fore and aft sections reconnected via two new solid bulkheads. The stern was reshaped, a large keel was formed to bridge the two assembled halves and Dad cast a solid lead base piece. After liberally waterproofing the interior we fitted a new deck (plank simulated in indian ink) formed mast and main boom from old disused broom handles, fitted out with brass rigging braces, again made by Dad, while Mum ran up some linen sails made from disused bed sheets. After a few coats of paint, from the outside at least, that model yacht looked every bit as good as some of the very expensive ones I had envied on the lake.

Admittedly, about twice as heavy as she should have been, the boat sailed, though not very fast, but no words of mine can convey any idea of the pride and sense of achievement we had. This model yacht remained at my home until the commencement of WW11 and the bombing of the street where I used to live.

The second friendly gift

Some twenty eight years later having moved to the Isle of Wight, I chanced to meet a young chandlers assistant friend of my son Gary. Knowing of the `gin palace' type of vessel his firm were apt to service, in the act of departure I had mischievously called over my shoulder "don't forget Barney next time you see a nice little boat for me...." Then, as I thought, taking up the joke, the young man called me back. Wishing I hadn't unintentionally misled him, I was taken aback when he said that in fact he had only recently obtained a nice little boat which might suit me. A little embarrassed I apologized about the joke and hoped I hadn't misled him. However he quickly reassured me and said he had obtained it as a going lot with one of the resplendent vessels shortly to come up for sale and the owner had said Barney could make what he liked on the little boat. Responding to my rather awkward hesitancy the young man said I could view the `little dinghy' on the quay having blue anti fouling paint on the bottom, and should I be interested I could have it for £65. I thanked Barney, returned home and told Irene I knew where I could get a very cheap dinghy.

For the fun of it, that evening we went along to the quay and were puzzled not to find any such dinghy. The only boat with a blue underseal was upside down and that was over twenty one feet, with centre board, stowed mast and rigging. We returned home disappointed and phoned Barney to check if we had the right part of the quay. He assured us we had and I shall always remember the slowly dawning sense of extraordinary good luck when I finally realized there had been no mistake and the little craft was ours, replete with thirty foot mast, bermudan rig, complete set of sails and a brightly coloured spinnaker. There was no catch in the purchase for I later met the builder of the boat, in the past it had won several round the island class races and was very, very fast! However there was only one snag, I had neither sailed nor crewed a full sized boat in all my life!

In at the deep end

We duly took the boat home, scrubbed, varnished and painted it till it looked like new with as fair a water line as any of the boats in Yarmouth, and called it Jinny after my mother. Some local people had warned me `to tuck the main sail in a bit', the time would come when I would find out!  One becalmed summer evening, just after we had put her in the water off a pontoon in the river Yar, I took my wife and twin daughters with the intention of `just having a look'. We had got aboard and I thought there is no harm in having a little paddle down the creek. However when only a hundred yards off the pontoon we had been interrupted by calls from a little cruiser chugging past. I instantly recognized the owner with Sherlock Holmes pipe hanging from his mouth as Brian G. one of the designers from the drawing office. We acknowledged his friendly wave with a cheerful `goodnight' and for a while heard the diminishing chugging from the engine of his boat. However I began to realize this was now getting louder. Again Brian overtook us and with another cheerful wave threw a line which I managed to catch, and with a thumbs up sign he continued to motor ahead towards the river!

Believing Brian had identified the situation and was kindly offering a short tow for us to row back, I was glad of this friendly assistance, but then I became alarmed when he continued on and out into the river. No doubt due to the noise from his engine, Brian seemed quite unaware of my attempts to attract him. Eventually he slowed up and turning, signalled for me to cast off and with another friendly wave and a `how's that suit you for size' thumbs up sign, opened up the throttle of his little cruiser and was on his way back. Later I would discover that on seeing us with sails furled and becalmed in the creek, he had towed us into the deep water of the basin `in order for us to catch a little wind'

By that time it was already getting dusk, there was a gentle breeze which could take us most of the way back to the creek from which point we could row. There was nothing else for it, I would have to set the sails and call on my scanty knowledge from those early model boat days. Having set the sails we began to run with the gentle breeze. This was easy, my family were nervous and privately so was I as we moved slowly back towards Yarmouth. The water was dead calm without a ripple, ahead it looked like a sheet of glass.

A boyhood dream come true. On the river Yar
with Jinny 1960

But then after a while I noticed a dark line stretched across the horizon. A dark line that seemed to me to be getting nearer and darker. Suddenly it was upon us, rain, lots and lots of pouring rain beating up the surface of the water. Looking forward from the tiller , torrents of it was cascading down off the now becalmed mainsail drenching us one by one. Then having got over the initial shock we sat helplessly resigned and laughing as if we hadn't a care in the world.

The boat now being heavier, the recommenced rowing was a bit tough, but we eventually made it back safely to the mooring, made fast, and stowed the sails. We were extremely wet, very cold, most relieved and made our way home. That was my first lesson on always being prepared for anything at sea and I was jolly glad for the time when as a youngster I built that model boat and learned the rudiments of sailing. But of no less relevance in the present context, when as a boy I dreamed and laboured on that tatty old salvaged model could I have believed that such longing for an out of reach experience would one day bear fruit by such convoluted means?

From sail boats to
hang gliders

When eventually I became reasonably proficient at solo sailing I often relived that first enforced trip to the basin and far beyond, tacking up stream before going about and enjoying the relaxation of the return leg, just as we had done that very first evening, running with the wind. With mainsail and spinnaker outstretched there were ample moments of reflection on the close affinity between sailing and flying. But perhaps it was inevitable that one day when the boat was out of the water and lying on its side, I reminisced on how much a Bermudan rigged boat looked like one half of a delta shaped wing. What I wondered, if a mirror image of the sail was placed on the other side? I had to find out.

I built a small model of jap tissue with balsa spars. Not only did it fly, but the chaps in the drawing office at lunchtime breaks enjoyed it immensely. With its light wing loading it sailed right across the office with the flattest glide we had ever seen!

Adjacent to our country home at that time were large fields with a steep gradient and when taking our dog Lassie for a walk, I often used to imagine hanging from such a contraption and sailing down the hill - just as early enthusiasts had done many years before - but I used to think `why not now?' Little did I know then that my double sided bermudan rig sailplane had been one of them! I then built a much larger outdoor model of some three feet span, using light weight aeromodelling fabric and spruce booms. When perfectly trimmed it flew down the sloping field for considerable distances as its low sinking speed matched the gradient. I was half way towards a full size machine which I could transport, unfurl, fly, and more to the point, I could afford!

A local councillor and businessman Mr.Spurr offered to sponsor it and take out patents which I was dubious about. Meanwhile a boat builder friend at the office built the spars and laminated front skid members. However, sadly Mr.Spurr died and as is often the way the whole thing took second place to other events. To my knowledge the spars and front skid still reside in my friend's attic.

Recently when visiting a young manufacturer to discuss my CM100 VTOL aircraft (shown later) I had cause to reflect a little. He is a leading world famous hang glider manufacturer and he started making them in bulk around 1984, today he is a multi-millionaire!

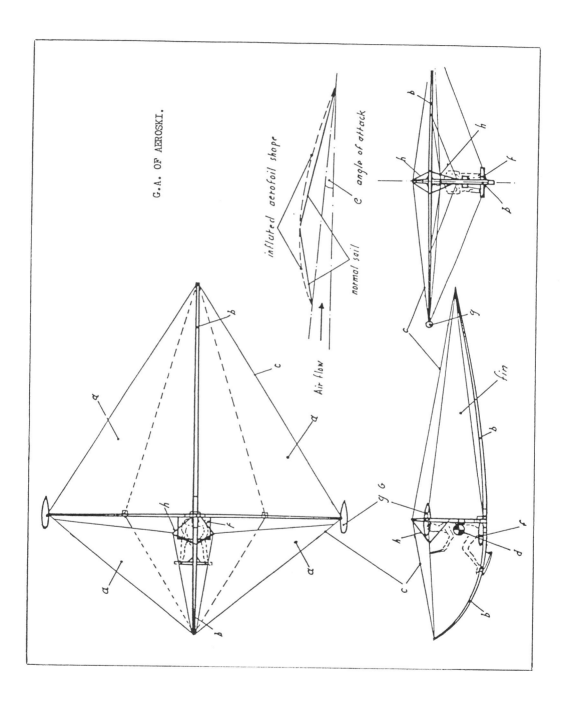

G.A. OF AEROSKI.

inflated aerofoil shape

angle of attack

normal sail

Air flow

fin

Original patent drawing for Aeroski 1960.

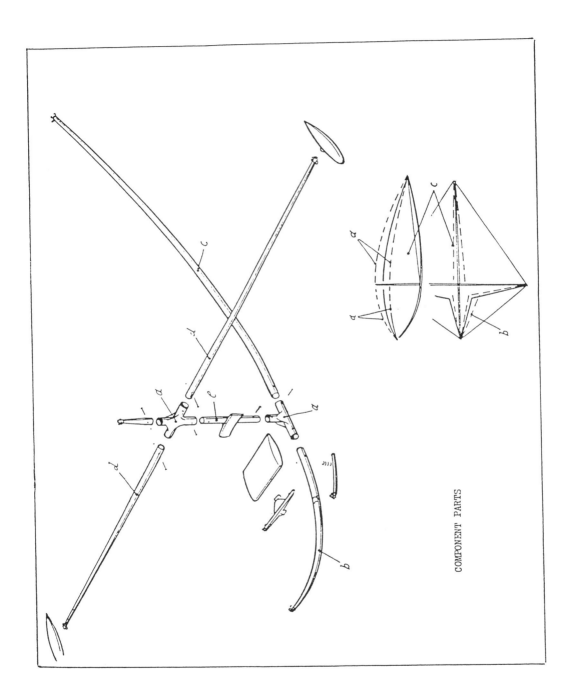

COMPONENT PARTS

Original patent sketches 1960.

FRIDAY,   AUGUST   17,   1962   SOUTHERN EVENING ECHO

ISLAND NEWSMAN'S NOTEBOOK

AERIAL SKIING IS IW MAN'S NEW SPORT IDEA

A NOVEL and exciting new sport, which, in the near future, could become internationally popular, is heralded by the latest invention of Mr. Leonard George Cramp, of Gardener's Cottage, West-lane, Norton, near Yarmouth.

Its name is aerial skiing, and instead of popping down to the beach next summer with your inflatable mattress under one arm, it may well be that in its place you will be carrying your own ski plane!

After spending a few minutes assembling the parts, you climb aboard and a motor launch tows you away for the start of an exciting and invigorating ride.

As the launch gathers speed the flexible wing above you fills and the water ski leaves the water—you're airborne. After soaring to a height of 50 or 60ft. you release the tow rope and from then on you are gliding.

PROTOTYPE

THIS is the sort of thing Mr. Cramp visualises and a business associate hopes, will be hap-

Mr. L. G. Cramp, of Norton, who has invented a revolutionary ski plane.

pening around the coasts, possibly quite soon.

It may sound far-fetched, but after watching a model of the ski-plane in action I can assure you it is not. However, within the next

couple of months, you should be able to decide for yourself.

Tests with models of the plane have proved most successful and work is going ahead on building a full size prototype. After land trials, it will be tested at sea.

The project is being backed by Mr. John Spurr, of Markeaton, Uplands-road, Freshwater, a businessman and member of the Isle of Wight Rural District Council.

Mr. Cramp's invention has been born from a lifelong desire to fly and to own his own sailplane. Conventional sailplanes, or gliders, are very costly, so for years he toyed with the idea of finding a novel and inexpensive means of flying in his own craft.

As he pointed out, the flying would probably be restricted but it would at least enable him to get airborne without having to join a club or "go without for a number of years."

PATENTS

HE explained: "Advent of the Ryan flexible wing stimulated thought, and suggested a means whereby the average chap could have a small glider of his own. The craft would be capable of supporting a person reasonably safely at low altitudes. This immediately suggested to me that the best application for such a machine would be over the water."

He added he had tried to produce a craft which was inexpensive, portable and lightweight, but which was controllable in free flight, like a glider, unlike other similar craft that have been built.

Until the design is protected by patents, naturally I cannot give full details of the ski-plane.

However, I can reveal that it is constructed of laminated plywood with nylon cloth wings. Tubular alloy may replace the plywood if the plane goes into production.

To me it looked completely unlike any orthodox sailplane or glider. Some people, I feel sure, will think it looks little more than a glorified kite.

Unless the numerous tests with models did not provide reliable information, which I think is most unlikely, then the ski-plane will enable a person to fly in safety for some distance.

Because the craft has a very low sinking speed, and is structurally light, Mr. Cramp believes it could also provide a novel form of sport from inland slopes.

SIMPLICITY

MR. CRAMP said the full size craft could be "picked up and carried under your arm. It would be less trouble than taking a small dinghy to sea and no more difficult to assemble or erect than a tent. The simplicity of this design could quite easily offer a new sport to the average man."

Mr. Spurr told me he was co-operating with Mr. Cramp on the commercial side and "in the event of the prototype proving successful we will proceed to put it into commercial production.

"I think it would be a fairly comfortable competitor to water skiing," he added.

When the prototype is completed it will be first tested by towing it behind a motor cycle.

Mr. Cramp is employed as a technical illustrator at Saunders-Roe. Before coming to the Island six years ago, he worked for the firm of D. Napier and Sons. All his interests stem from anything that flies.

FLYING SAUCERS

A FEW years ago he gained considerable publicity by designing a flying saucer type aircraft.

He has done considerable research into flying saucer phenomena and means of gravitational propulsion. He has lectured considerably on "gravitics" and is the author of the book "Space, Gravity and the Flying Saucer."

In a room at his home hangs a magnificent oil painting of a flying saucer that was seen in the Island last year. He interviewed the people who saw the machine and from the description given to him painted the canvas.

Much of his spare time is taken up by writing technical articles for such publications as "Aeronautics." He designed a turbine 'ram jet engine and for many years was interested in ground effect machines, similar in principle to the Hovercraft.

One of his designs in model form appeared in the "Modeller" and another was on view at one of the model engineering exhibitions in Newport.

Copy of the original newspaper item in 1962. Although at that time the american Ryan flexible wing had been proposed for space craft recovery, hang gliders as such had not yet been introduced.

1976 - 1992 The Skimmer. The fully operational one tenth scale model was demonstrated to BP engineers on the Isle of Wight and later at their R&D establishment at Sunbury in 1978.

ATF 16 years

The Skimmer was designed to skim the water in surface effect by exactly the same principle as that in the `new secret' Russian vehicle revealed in the `Today' news item of June 5th 1992, but Skimmer - unlike the `Aquatain' - by vectoring the forward propulsion units, could hover like the Hoverplane (shown later) at zero forward speed, while out of surface effect it could fly like an aircraft over any surface.

The full size prototype would have skimmed at a height of fifty feet with a competitive fuel consumption of only .046 lbs passenger mile at a useful speed of 120 mph.

This Skimmer model was later displayed *publicly* at an *international* convention in Brighton, UK, by an over indulgent sponsor without the author's full notification. Indeed by now I was becoming accustomed to being treated more like a workshop foreman by some of these people. I regretfully found it common practice for `sponsors' to bring unannounced financially interested parties, who having been treated to a demonstration and know-how speech by me, were drawn aside by the sponsor to go into huddled, whispered conversation about *my* project!

Neither were `away' demonstrations always a good idea, as I had cause to reflect when giving such to an assembled team of sixty high ranking officials at their R & D Establishment at Sunbury, Surrey,UK. In 1978. The R & D Hall is vast, some sixty feet or so high, ideal for flying models.

As will be seen fron the following photographs the Skimmer scale model was over six feet long, electrically powered via umbilical cables and was radio controlled. Prior to this after lunch meeting my son Gary and I had set out the cables and radio and check flown the model, the lift offs from the huge polished floor were very impressive. Then Gary had to leave the rest to me as he had another appointment. After lunch the hall began to echo with the voices of the assembling officials and dignitaries examining the Skimmer model on its stand. After due introductions and brief description given by Professor Desti, Dr.Johnson and myself, I started the propulsion motors. As always Skimmer gracefully surrendered its surface contact and went into hovering mode.

First, as a warm up, I did a few slow circuits in surface effect, then I opened the throttles for STOVL (short take off and vertical landing). Skimmer immediately responded by going into a most undignified horrendous stall and ungainly dropped back into the cushion mode. Midst apologies and audible mutterings concerning `Murphy's Law', I checked everything and tried again, and yet again, without success. Despite my embarrassed apologies and courteous reassurances from my assembled audience theywere becoming increasingly impatient as indicated by occasional reference to their wristwatches, so there was little for it but to cancel the meeting.

To lay on such a demonstration involving transport of the model and accessories from the Isle of Wight to Sunbury and make arrangements with such a large group of very expensive people, is not something achieved easily or very often, and so , sadly, Skimmer was never seen behaving in all its impressive glory at Sunbury again. A few weeks later we established the half suspected cause. In such an establishment there is much laboratory work going on, often involving radio transmitters, and Skimmer had been smitten with R/C interference!

The press release on the Aquatain coincided with the preparation of this section and it is interesting to note the reporters `off the cuff' descriptive term for it (*The Hoverplane*) although this is of course most likely to have been merely unfortunate journalistic licence.

R/C Skimmer model undergoing static hovering tests in water tank.

Workshop photo of Skimmer. Note the angle of forward propulsion fans directing plumes beneath side plenum chambers. Compare with `Aquatain'.

Skimmer basic main frame before the addition of plenum side walls, rear end plate, ducts and flaps.

Assembly of components including side-walls, end plate, fin and forward swivelling ducts.

Construction of propulsion ducts. This method of fabrication was long before the availability of model airctaft ducts.

Skimmer investigative design changes. In this particular instance, the author's twin cascade principle was used, later to be employed in the TSA vehicle as shown in Chapter 8.

Left. Preparing Skimmer for first trials with propulsion fans, motors and radio installed prior to final skinning.

Right. Skimmer tethered in hovering surface effect at four inches above the bench.

Left. Out of surface effect in static tethered hover mode at four and a half feet. Note, the structure beneath the model is crash preventative rather than supportive.

Right. In surface effect over water. Note the swivelling ducted fans injecting air downward and rearward to provide cushion lift in the side plenums, together with the anhedral shaped underbody of the craft when in forward motion.

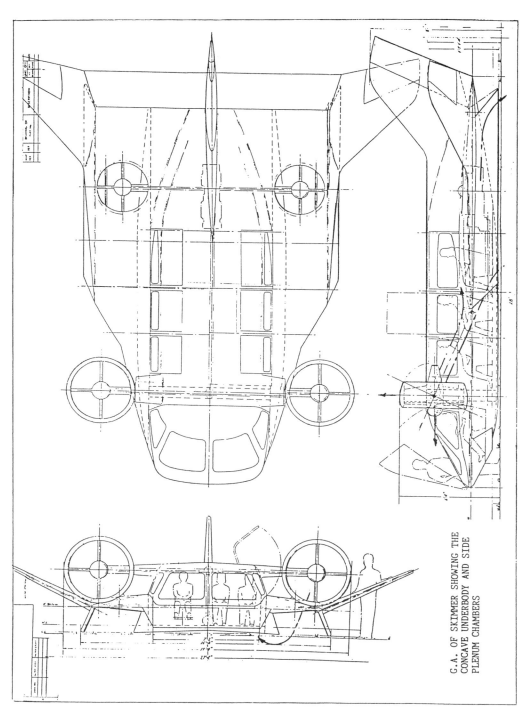

General arrangement of the Skimmer. The R/C model was being demonstrated before sixty officials at B.P. R/D establishment at Sunbury in 1978 when the operation was cancelled due to extraneous radio interference.

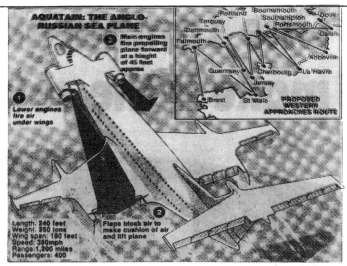

The plane speeds about 40 ft above the sea at more than 300 mph

The top secret prototype could open up to carry 400 commandos

Hop, skim and jump to France

by ELLIS PLAICE
Aviation Correspondent

A RUSSIAN secret weapon designed for a Cold War invasion of Britain could now be used to whisk holidaymakers off to the sun.

The giant Aquatain hover plane could cut cross-Channel journeys to minutes by skimming waves at more than 300 mph and hopping over any ships in its path.

Talks have begun between the Civil Aviation Authority, Lloyd's insurance and City financiers, TODAY can reveal.

Consultant Stewart Lawson, who is negotiating with the Russians, said: "This is no flight of fancy. The prototype is already flying, and production is to begin next year."

The plane was designed to skim in below radar defences and unload 400 fully-armed commandos on to an enemy coast.

Instead of using conventional wing "lift", engines fire downwards and backwards, building pressure beneath a screen of flaps.

Main engines then propel the plane on a "bubble" of air about 40 ft above land or water.

Russia plans public trials of the £6 million Aquatain in the Pacific next year.

'Today' June 5ᵗʰ 1992 press release of the 'Aquatain secret breakthrough' skimmer type craft. Close comparison between this and the author's machine of 1976 reveals the identical surface effect and aerodynamic system.

7
Models to Supercraft

1943 - 1958 ATF 15 years

As is shown by the author's published articles and plans on SEVs (Surface Effect Vehicles) this work preceded the eventual patents by Sir Christopher Cockerell by over a decade. Much of these efforts were instigated to help pay the bills. It will also be apparent that it stems from the pulsed coal gas combustion chamber experiments at Hatfield in 1942. Indeed, the memory of that fascinating 'levitation' experience was never far from my mind. Thus it was that I finally got round to exchanging the small pressure rise produced by controlled high frequency explosions for a continuous pressure rise produced by an axial flow fan, and for me the open plenum type hovercraft was born. Later I would learn that the idea was by no means new and I was just one of many who had found it. Even so, for a while I enjoyed the excitement, and continued building and experimenting with models, some of which would later be used in articles and sold as kit plans. However the income from these didn't even cover the material costs of the models, but every little helps! Concerning the hovercraft there exists assumptions which in fairness to others I feel should be put right.

It is now generally assumed in Britain - and for that matter all around the world - that Sir Christopher Cockerell *invented* the surface effect (hovercraft) *principle* per se. There can be little doubt that he would be the first to agree, he did not. He may not have been any more aware of the fact than this author - and many others - that the *idea* of air lubricated skimmer craft was recorded as far back as 1716, as will be seen in the Appendix to this volume. However he did patent a surface effect vehicle with a circumferential or *annular air jet* system, which in some circumstances can considerably augment the jet lift by trapping a cushion of pressurized air which is formed beneath *all* concave shaped air lubricated (or surface effect) vehicles, termed plenum chamber craft. Moreover I have good reason to believe a friend of mine, Marcus C, who had more experience in publicity than engineering (employed at the time in the technical publications department of BHC) was influential in the title 'Hovercraft'. I pointed out to him the technical inaccuracy, for all such machines still record weight even when operating over scales. The *true* hovercraft is the helicopter and air supported vehicles (ASVs) like it. Even so, the name took on and in this sense the choice proved correct.

When on rare occasions - no doubt due to a minister's reaction to popular appeal - officialdom *does* respond to innovation, even then they can't get it right and are apt to 'go over the top' in their ill founded enthusiasm. Allowing that Cockerell's masterly invention could nonetheless be designated an *improvement* to an already *existing* idea, in certain circumstances some engineers question whether the extra weight together with the increased aerodynamic duct losses - inherent with the annular jet principle - are even worth it. Indeed some have said the situation is directly comparable to an aerodynamicist *improving* the efficiency of aircraft wings, as did Fowler's well known flap, which has been largely responsible for the successful application of air transport as we know it today. True there are very few of us who would refuse the accolade that Cockerell received, but many people `in the trade' feel the magnitude of the award was out of all proportion to the contribution made.

Not only had air cushion machines been built and operated all over the world, Russia, Sweden, Canada, Britain, America and Australia, but in fact at one time there was a very well known vacuum cleaner called the `Constellation', which some readers may remember, employed the air cushion system, and as stated, the author had experimented with it years before. But a little story concerning this may serve to illustrate the main issue involved here.

For several years a good friend of mine Ken D., being an expert photographer, used to come to my workshop and occasionally take photos of the models for me. One such model was circular in planform and painted orange, I called it a `skimmer' he called it, not disrespectfully, `that thing'. At that time the Saunders Roe company was facing problems and had circulated requests for any ideas from staff. Ken suggested it might be useful to show the firm this model. Unfortunately I had to reject this, because I was already committed to a large shipping company on the Clyde in Scotland, who were showing keen interest in the concept. However due to reasons expressed elsewhere in this book, no patents had been applied for. One day, several years later, Ken D. chided me for not telling him that I had in fact offered the design to the firm. Having got over my surprise and denial of this, he said he had just returned from No.2 wind tunnel and seen `my' 2ft. dia. model there. He was so certain it was the same model he had photographed for me, he thought I was `being cagey'. So much so, I had some difficulty in trying to convince him that I knew absolutely nothing of what he was saying. He ended the rather hot discussion by recommending that I go look for myself. The rest of the story is rather academic. I did go to No.2 wind tunnel and saw there a 2ft. dia. *orange* model. To my familiar eye there was sufficient difference, but that model was Sir Christopher Cockerell's. How hard can fate be!

It is interesting to note there were two other prophetic incidents concerning this model which occurred in the early 1950s. The first when I was demonstrating it to the directors of the above mentioned Clyde company on the flat roof of a Southampton hotel. I had been dismayed when after an otherwise successful demonstration the fan disintegrated. Retrieving the model I glanced up to note an accompanying naval architect in the act of furtively tucking something into his waistcoat pocket. The man grinned sheepishly at me, then proffering one of the little fan blades he said "You never know, there might be a small piece of history here."

The second incident occurred when `flying' the model on our lawn in the presence of my elder brother Alfred. He had noticed that the fan made a neat swarthe in the too long grass and said, "Wouldn't that make a smashing lawn mower." Considering the end product I had in mind, I barely noticed what I thought was a light hearted joke. However, in view of the millions of hover mowers throughout the world today......! I still have that little orange model in my workshop loft, dusty and timeworn and when visitors spot it and ask `what is *that?'* more often than not I am tempted to tell them, then respect for our mutual available time prevails and I side track and pass it off. I had no idea then that the general principle had been internationally known for many years. I was to discover that even the 18th century mystic and scientist Emmanuel Swedenborg had a design which was to be *manpowered* by paddles. If only he had had an engine!

The phenomenon of ground effect is well known to aircraft engineers. In fact every aeroplane or helicopter that ever flew experiences a degree of so-called cushion effect when operating near the ground or surface, whatever that may be. Many inventors tried to exploit the idea, some of them well known, many unknown. One of these, although historically well known, his interest in hovercraft or surface skimmers, certainly was not. `He' being none other than Lawrence of Arabia, who with the help of an aircraft technician, built and tested a piloted version!

Early designers independently pursued similar paths to produce a vehicle which would operate off road and skim along at significant air gaps, (height above interface). But with the introduction of Cockerell's brilliant peripheral jet system and consequent increased research, it soon became evident that the off road, over hedges concept was going to be neglected due to an inherent limitation in the idea. That is, the higher a surface effect machine hovers, the more power it requires, until a point is reached where the craft is operating outside surface effect benefit and therefore requires much the same power loading as the helicopter. In other words, the closer to the surface a hovercraft `flies', the more efficient it becomes. But an additional problem other than efficiency is net at increased height, that of inherent instability. Engineers discovered this becomes a serious problem when the machine operates at heights divided by the craft effective dia. $[\frac{h}{de}]$ equals more than ·2 . Therefore these two reasons alone were sufficient to encourage designers to look for alternative solutions, including the present author.

Now it is obvious if you operate such a machine too close to a rough surface it is likely to sustain damage and that is exactly what happened to the first Saunders Roe SRN1 hovercraft. However as shown in this section, several years *before* Cockerell's invention, my eldest son solved the same problem with a small experimental model by wrapping a strip of flexible tape around the bottom periphery. After offering some well meaning tentative compliments, I, in my questionable wisdom, pointed out to him "But you can't do that son, it's cheating". That simple piece of tape was probably the world's first single membrane flexible skirt and *several years* ahead of its time. After some eighteen months design, rebuild and further SRN1 damage, it was eventually introduced by Cockerell as a new idea. More significantly without those skirts the hovercraft could never have advanced far beyond the stage of yet another interesting idea, with or without Cockerell's peripheral jet.

From dreams in 1948 to actuality, the following depicts the author's bright orange coloured model hovercraft dubbed by the editor of the magazine Model Maker `The Hoverer'. Although some of the more philosophically inclined might assure us that such experience `is good for the soul' it is, to say the least, rather irksome, having spent years labouring on one's idea, to witness workmates and scoffers becoming elevated to relatively highly paid positions around you working on it. Not helped by the intentionally kind remarks of friends and neighbours `but didn't *you* invent that?'

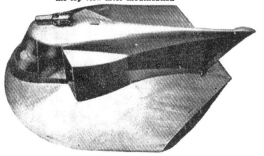

JULY, 1960

Model Hover Craft

An account of experiments including a practical model

By Leonard G. Cramp, A.R.Ae.S., M.S.I.A.

THE increasing public interest in hovercraft type vehicles has encouraged the writer to offer this brief account of some work carried out a number of years ago.

It will be seen that the original ideas pursued then were, in many respects, identical with present trends of development, moreover it must be stressed that this research was of a private nature and does not intentionally represent existing designs.

The principle of hovercraft or air cushion borne vehicles is now generally understood, but a brief word or two as an introduction will not be amiss. It is basically a means of differentiating air pressure above and below a pressure plane, but—unlike an aeroplane wing—there is no reduction of atmospheric pressure above the plane to any marked degree. Fig. 1A shows such a plane where it will be seen the supporting surface or ground is in

explosions. Such a device is subject to excessive vibration, but it served to prove the point.

In operation it is similar in principle to the so-called plenum chamber type craft now being investigated in Great Britain, Switzerland and the United States of America, Fig. 1d.

The writer's conception of a forced air operated machine on these lines was approached from a slightly different aspect than that of a pure plenum chamber; the more developed theory is shown in Fig. 2a. Air was passed through a propeller as in the plenum chamber Fig. 1d, escaping to atmosphere via the periphery as in the previous case. But the underside of the craft was so designed as to direct the airflow smoothly downwards towards the supporting surface where it curved away towards the rim of the craft. Thus it will be seen from Fig. 2a that there is a relatively large mass of trapped air

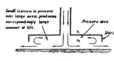
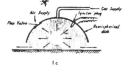
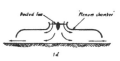

Fig 1a

1b

1c

1d

fact one of two restraining walls trying to prevent the incoming air from escaping, while the second wall is formed by the plane itself.

It will be appreciated that the aperture through which the expanding air escapes has a critical function, its area increasing constantly with increase in pressure plane size. Theoretical available work is completed when the air pressure is expanded down to the surrounding atmospheric pressure at the exit.

The natural development of the idea is to form the plane with a skirt as in Fig. 1b, in which it will be seen that some of the kinetic energy in the outflowing stream is again converted into pressure energy through turbulence.

A conducted experiment embodying this basic principle is illustrated in Fig. 1c. In this the pressure plane was formed by an inverted dish shaped bowl into which provision was made for a constant supply of coal gas. A simple flap valve regulated air induction, whilst a small glow plug was employed for ignition.

In effect, the whole formed a simple combustion chamber, in which a mixture of gas and air could be ignited. The operation is almost self explanatory. First the igniter was lit, then gas was allowed to flow into the chamber. As soon as a detonating mixture was reached, the charge ignited, the resulting increase in pressure forcing the dish upwards. The instant the dish left the supporting table, the gases escaped, the ensuing suction bringing in a fresh charge, and the process continued.

In effect it will be seen the principle was identical to the pulse jet cycle employed in the VI during the war.

By this means the dish was kept "airborne" by the increase in pressure caused by the high frequency

between the pressure plane and the outgoing stream and it will be apparent that due to viscosity this mass of air will be induced to revolve in the fashion of a toroidal belt, its velocity approaching V—that of the outgoing stream. In practice this was found to have been in the order of 90 per cent. of V and therefore represented quite a usable amount of energy.

The manner in which this energy is converted into lift is quite simple and calculations were based on the following basic assumptions. The revolving mass of air exerts an upward thrust on the underside of the pressure plane due to centrifugal force, whilst the outward moving stream suffers a downward but equal centrifugal "pinch"

Model above is as shown on G.A. opposite. Below is the top view after modification

---

* This experiment was first conducted at Hatfield, Hertfordshire in 1944 and the resulting 'Hoverer' model bore such a remarkable resemblance - including finishing colour -to the model built for Sir Christopher Cockerell, that it caused confusion at the Saunders Roe, Isle of Wight, test tanks when his model was later tested.

Bottom view of the 'Hoverer' model showing plenum, circumferential buoyancy cell and anti-torque blades.

"Wouldn't that make a smashing lawn mower". 'Hoverer' leaving a swath through long grass. The little girl in the top right of picture is the author's twin daughter Sue, now a mature mother with a family of her own.

*One of the first stable powered 'flights' of the bright **orange** coloured 'Hoverer'.*

Checking the hover height with the help of son David using equipment from my parents new bungalow as mentioned in Chapter 5.

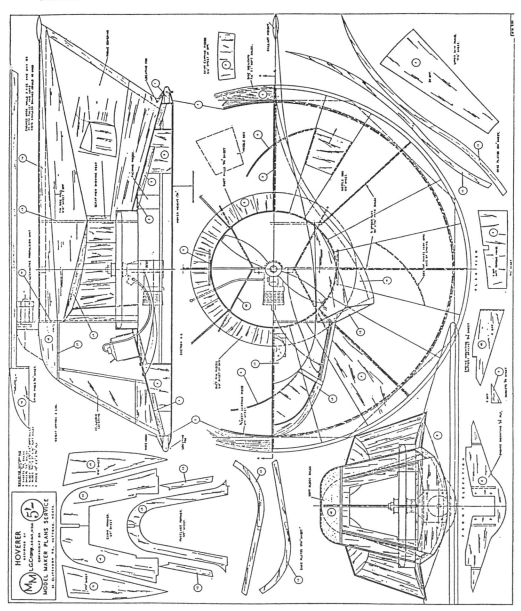

The 'Hoverer' kit plan, the author's 'Skimmer' dubbed 'Hoverer' by the editor.

MODEL MAKER

ONE of the problems associated with the hover-type vehicle is that of stability, for as the craft is literally balanced on a film of air, a small change of weight will tilt it This in itself is no serious dis-advantage, but an accompanying phenomenon presents a greater problem. It is the fact that due to the tilt more air is spilled from the higher side than the lower and consequently the craft experiences a small thrust in the direction of the lower end due to the increased mass reaction.

On a full-size vehicle this problem can be offset in a variety of ways, by the use of deflector blades, thrust ducts, and the like, but on a smaller craft this approach may not be the best alternative. Again there is the possibility that the vehicle will tilt due

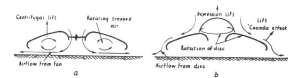

Fig. 1

ORBIT

A spin-stabilised hover vehicle for electric power

Designed by

L. G. CRAMP A.R.Ae.S., M.S.I.A.

to a loss of chamber pressure, as would be the case if a large indentation in the supporting surface were encountered.

The small model described here was born as a result of the author's attempts to find a possible solution to the problem. As an alternative idea it may have some merit, though it is possible that it has inherent disadvantages which render the idea imprac-ticable for full-size practice.   Nevertheless it does make an interesting little model and is well worth the work involved.

As will be seen from the accompanying photo-graphs, it rather looks like the current descriptions of the so-called Flying Saucer, which together with the fact that it operates while airborne makes the idea even more intriguing.

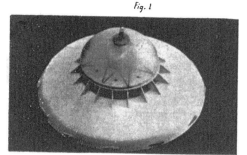

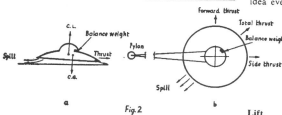

Fig. 2

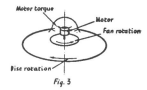

Fig. 3

Lift

By now those readers who have followed these articles will be familiar with the pressure plane or air cushion principle of lift, but as the spin stabilised model lends itself to an additional aerodynamic effect, a brief description of the combined principle of operation may be necessary.

The author's conception of an air cushion borne craft is based on the fact that air directed down-ward from a ducted fan is forced to bend outwards by the supporting surface or ground and if the under-side of the vehicle is suitably designed, the relatively large amount of air trapped between the outgoing stream and the underside of the craft, will be forced to rotate due to viscosity, Fig. 1a.

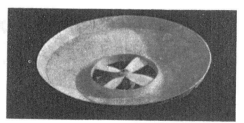

Probably the first hovercraft to be fitted with flexible skirts; a strip of masking tape suggested by the author's elder son Gary.

OCTOBER, 1960

If the plane form of the craft is made circular, with a suitable designed cross section, and then made to rotate at speed, two additional effects are consequently manifest. First the craft is stabilised through spin and secondly the upper surface of the rotating disc experiences an additional and useful aerodynamic lift.

This is accounted for in the following manner. Again due to viscosity, the air in contact with the whirling disc is thrown outwards by centrifugal force. This induces a flow in a radial direction over the section of the disc which causes an aerodynamic reaction, more technically recognised as the "Coanda effect" after its discoverer. See Fig. 1b.

shaped cabin to the power input wires, this being the simplest and most obvious way to achieve the same result.

Propulsion

As it was considered that the employment of propulsive jets for forward motion might make the model too complex, the author has relied principally on thrust from the spill effect mentioned earlier.

This is quite simply achieved by off-setting the centre of gravity a little. A small piece of Plasticine placed inside the rim of the dome proved sufficient with the author's model.

Fig. 2a shows how this same principle serves to keep the model in forward motion, while at the same

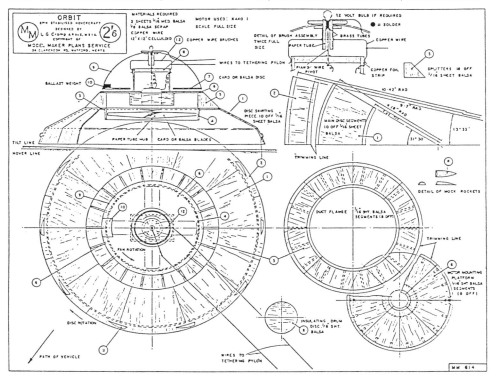

In addition to this effect, there is a depression formed over the disc and dome-shaped cabin, again due to the centrifugal force of the air, this in turn is more commonly recognised as a vortex or whirlpool. So that, in theory, the energy lost through friction in rotating the disc in order to balance it is compensated for by the additional lift. It will be appreciated that while these lift effects on such a diminutive model are probably negligible, the performance more than justifies its construction.

Naturally, in the case of a full-scale version, the cabin would be fixed and in order to maintain realism in the model, it was decided to "anchor" the dome-

time it can be made to move outwards away from its tethering pylon should this form of operation be preferred. Fig. 2b.

Construction

On studying the drawing, it will be seen that the main disc is completely fabricated from balsa segments, the whole being assembled without frame or structure of any kind. The reason for this is, of course, that it is inherently robust enough to render a structure unnecessary.

The centre fan duct should be constructed first from $\frac{1}{4}$ in. deep 1/16 in. sheet balsa with the grain running across the width. As with August's model

The 'Orbit' kit plan.

AUGUST 1960

BUZZIN' BEE

A simply-built

electrically operated

hovercar by

LEONARD G. CRAMP

A.R.Ae.S., M.S.I.A

THE principle of lift by means of a pressure plane or air cushion and its accompanying advantages lends itself admirably to road-borne traffic and although there are many basic problems to be overcome before this type of transport becomes a reality, a great deal of research is already under way. The Curtiss Wright "air car", recently covered by the press and television, serves as a good indication of the shape of cars to come.

Apart from the fact that this type of vehicle is amphibious and would not therefore be restricted to a roadway, the structural advantages from a car designer's point of view are many. Foremost among these is, no doubt, the completely revolutionary form of suspension, for the hovercar and its occupants literally ride on the ideal shock resisting fluid — air — which is available without the requirement of recoil tubes and pistons. This implies that the whole basis about which the modern car is designed, wheels, suspension, chassis, engine and body, is reversed to, body, engine, lifting and propulsion system.

The average complete car wheel weighs something like 25 lb., which means a weight penalty of 125 lb. for most cars. Add 200 lb. for the front and rear suspension and we have 325 lb. dead weight which will not be required with the hovercar. Even at the present time the most modern car is a paradox of engineering development. On the one hand we have the beautifully-finished exterior which is intended to "look attractive to the eye" in addition to being functional, while on the other hand, the underside of even the most modern car is a filthy sight, after only a few weeks of running.

The suspension and mechanical components require constant lubrication and for that purpose greasing nipples are provided, the ends of which eventually become clogged with fine grit from the roads, so that some unscrupulous garage lad, neglecting to clean the nipples properly first, injects a mixture of high pressure oil and grit, which, of course, is a very efficient abrasive. Bearing all this in mind, it makes one wonder how the mechanical parts of a car last as long as they do.

None of this need apply to the hovercar. In fact, if properly designed, the underside of the vehicle will be just as presentable as the exterior and hosing down will be quite a pleasure.

On the debit side, there is the possibility that controlling a hovercar might present quite a problem when "airborne", for it does after all experience far less friction than a conventional vehicle — for example, skidding on ice. This problem and others accompanying it are being currently investigated and most probably they will be overcome.

Such small experimental models as the one described in these pages are helping to solve the "teething problems" of the hovercar, though it will be appreciated that this one is far from being an integrated design. Nevertheless, such a model can be a very instructive introduction to the pressure plane lifting principle.

Lift

On studying the plan, the reader will notice that the underside of the model has not been "cleaned up", *i.e.*, the bulkheads have been left exposed to the air flow. This is intentional and makes very little difference to the performance of the model; a glance at *Fig. 1b* will help make this clear.

A brief description in the previous article dealt with the principle of lift and showed how the airflow from the fan was directed downwards and then bent outwards by the supporting surface. The trapped air in the surrounding space is thus induced to move with the outgoing airstream and consequently caused to revolve. This imparts a centrifugal "pinch" on the outgoing stream which tends to prevent it from escaping, whilst the pressure plane of the vehicle experiences a centrifugal reaction upwards, which together with the increase in pressure thus formed, generates the lift (*Fig. 1a*).

Similarly, the bulkheads in the model hovercar cause an obstruction and therefore form separate chambers in which the trapped air rotates. Although the net result is much the same, a full scale counterpart would be more carefully finished, complete with buoyancy chambers, etc.

Propulsion

It will be seen that due to the forward tilt of the motor, there is a very slight thrust component on the car, which contributes to the total thrust, and while this arrangement also offers more clearance for the motor, the airflow through the fan is encouraged into a rearward direction towards the propelling nozzles, which form the main propulsive jet.

Twin cabins giving the appearance of insect eyes and the noise of operation make "Buzzin' Bee" an apt name for this unusual model

Beginning of the Supercar. Compare with Chapter 12. This model was demonstrated at fetes, schools and model engineering exhibitions long before this article was published. 'Buzzin Bee' was a label given by the author's eight year old son David.

AUGUST, 1960

Construction

Commence by cutting out and completing the bulkheads which should be set aside to dry. Note that bulkheads Nos. 1 and 7 are cut from ⅛-in. sheet. The front and rear propelling grilles can now be cut out and assembled on to bulkheads 1 and 7. Next form the nose and tail baseplates from ⅛-in. sheet and pin them carefully into posi.ion on the plan. The ¼-in. square balsa side rails can now be added and secured with pins.

The bulkheads should now be cemented into place and, while drying, the front and rear spine pieces cut from 1/16-in. sheet. Position these in place and leave to set.

As it is necessary to construct the fan duct accurately, the following procedure is recommended. On to a sheet of fairly thick card draw a 3⅜-in. diameter circle divided into four equal parts by two lines through the centre. The disc should then be cut from the card and pinned down on to the work bench with a drawing pin pushed through the centre hole. Next cut several 3-in. lengths of balsa ⅜-in. deep with the grain across the width. The author has found that a smear of cement applied to the inside face of the duct promotes natural bending due to the cement shrinkage and apart from the added strength saves hot steaming. The lengths of balsa should now be fitted one at a time around the card disc, cemented together and held in place by pins pushed into the workbench. When set the pins can be withdrawn and the duct ring and centre disc removed from the bench. The whole can now be positioned into the framework between bulkheads 3 and 5, making sure to line up one of the centre lines on the disc with the two spine pieces and allowing 1/16-in. clearance for the skin thickness. The two halves of bulkhead 4 can now be set in place to the remaining disc centre line, cemented and left to dry.

Planking is fairly straightforward, using wider 1/16-in. sheet pieces around the duct, and narrow strips around the sharper bends. The author found it easier to lay the wider strips temporarily in position and to make the outline of the duct wall with a pair of compasses, using the card disc centre pin hole as a pivot.

The two end panels below the grilles should now be filled in with 1/16-in. sheet, with the grain running vertically and the whole body sanded to a smooth finish. The duct card disc can now be removed and the motor mounting fitted, but as the reader may wish to install a different make of motor, the type of moun ing is largely a matter of choice. It must be borne in mind, however, that for maximum efficiency the least pos:ib'e obstruction across the duct is desirable.

The fan, which can be quite easily made from balsa sheet or even thin card, can now be constructed and fitted to the motor shaft, leaving a minimum clearance at the blade tips to allow for movement of the fan due to vibration.

The small rudder can now be fitted, together with the body cockpit covers, fins and other finishing details,

the fan duct grille being optional. The whole model can be covered with thin tissue — depending on the finish desired — and painted.

Operating

The model should not require balancing, but should it be necessary to do so, the addition of a small amount of plasticine fitted to the underframe at the extremities should be sufficient.

Some experimenting with the fan blade angle might be

(continued on page 395)

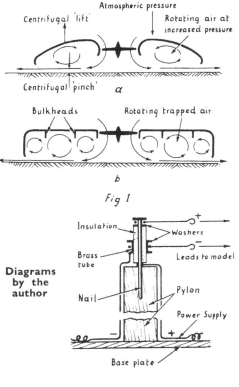

Diagrams by the author

Fig 1

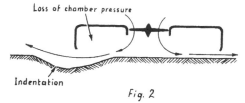

Fig. 2

Fig 3

'Buzzin Bee'

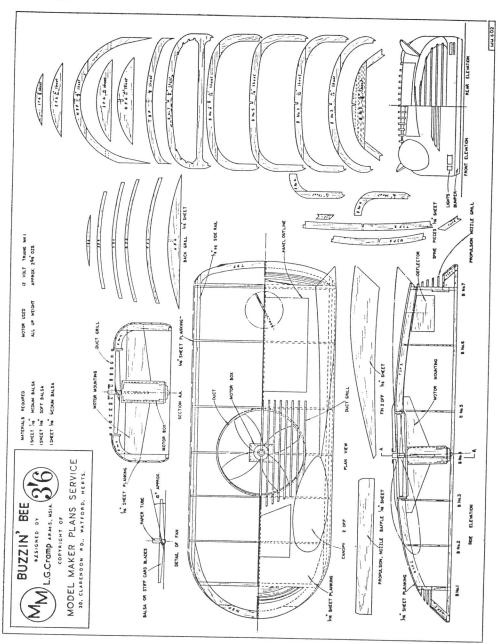

'Buzzin Bee' kit plan.

This photo shows basic frame and single central fan duct.

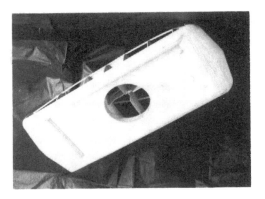

This bottom view shows the typical plenum and duct with circumferential buoyancy cells.

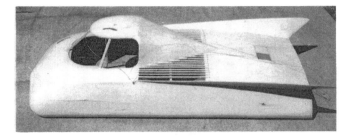

Finished model was over three feet long and was electrically powered. Hover height was over one inch, top speed twenty miles per hour.

One of the experimental arrangements conducted by the author in which side wall slats were used, similar to aircraft wing leading edge slats.

The natural development of the circular 'Hoverer' published in the Model Maker 1956.

Cover of the magazine which carried the article of the 'Hoverer' model demonstrated at Southampton in 1956.

1955 - 1963 Investigations into disc aero-forms with surface effect.
ATF 8 years.

This work was primarily initiated in order to exploit the author's ideas for a minimum structure - and therefore light weight - flying wing type aircraft. It is interesting to note that even the terminology employed by the Russians later version in 1963, was word for word identical.   The late Major Oliver Stewart, the original commentator for the Farnborough Air Show, UK, and editor of the popular magazine `Aeronautics', visited the author's workshop in 1957 and showed keen interest. It was chiefly due to his motivation that led to the article on disc wing aircraft shown in this section, the philosophy of which I incorporated in many of my later designs.

With due respectful acknowledgement, over recent years it has been amusing to me to note that some Ufologists are still quick to identify a clandestine link between the Canadian Avrocar, also listed in this section, and other similar aircraft, with anti gravity research now being conducted in the USA, Russia, and other parts of the world. However in so far as I have a more than peripheral identity with the basic design of some of these (beyond a convenient somewhat obvious silent official response concerning this) I can say there is absolutely no truth in this rumour. It seems to some, *any* aircraft of a circular or near circular planform configuration, just has to be linked with extraterrestrial flying saucers. Indeed in 1956 I went to some length to explain in Piece for a Jigsaw that there are considerations other than aerodynamic ones which determine the circular shape of *gravitationally* manipulated space craft. In this report I hope the reproduction here of the relevant part of my article on aerodynamic disc wings, published in *1958* and later in Piece for a Jigsaw in *1966* may help to clarify this issue a little further.

The idea first came to me when many years ago I used a cycle as my only means of transport. I often reflected on the fact that the bulk of my weight was transferred to the wheels by the upper spokes in tension, rather than the lower spokes in compression. This became even more obvious to me when I read that some early builders of very light aircraft in order to save weight had in fact used canvas covered *string* spokes in their wheels.! Indeed my first biconic section disc wing comprised merel a balsa wood rim connected to the centre hub by nothing more than an upper and lower tensioned Jap tissue skin! It weighed only a couple of ounces and when ballasted for forward C.G. balance and fitted with a couple of fins, it had a very flat glide.

As will be seen from the following article, when I later realized such a planform was the ideal requisite for utilizing surface effect, for me the disc shaped aircraft was born. My enthusiasm for it was based entirely on light weight and aerodynamic principles and had no relation to what were to become known as flying saucers.

Exploration of the
surface effect

As is now well known, all the early hovercraft inventors employed the upturned dish principle for their craft, later termed plenum chamber, in the middle of which was situated a fan. This pumped air into the plenum chamber at increased pressure until the resulting lift equalled the weight thereby causing it to rise, thus for a given weight, or greater pressure, the higher the rise will be. However requiring VTOL as I was then, I became one of the first to discover that if such a small dish was hovering on a cushion of air, and one tried to physically assist by lifting it, a negative lift, or suction was experienced, this in keeping with the well known Bernoulli's Theorum. This can be explained by the fact that when the vertically descending air stream from the fan meets the surface over which the dish is hovering, it is turned through a ninety degree angle and spread out through three hundred and sixty degrees. The outward radiating air stream between the dish and the surface and the accompanying pressure drop, is the cause of this negative lift. Now by directing the air from the fan, via internal ducting, towards the periphery of the dish and ejecting it downward through a circumferential slot, as patented by Sir Christopher Cockerell, the situation is reversed, in that the foregoing pressure drop, is now manifest as pressure rise. Furthermore this pressure rise can be maintained with increased height.

However there is a penalty to pay, in that increased height requires increased power, therefore the hovercraft principle as it has now become is a close surface effect phenomenon, which can only be successfully employed by the use of deformable flexible members we now know as skirts. In addition all the lift air has to be ingested into the body of the vehicle then directed to the perimeter jet, before being ejected beneath the craft, and this ducting implies not only weight, space, and high duct losses, but increased power. All of which is too severe a penalty for realistic VTOL aircraft.

I reasoned that in theory much the same lift could be generated without some of these penalties, by placing a series of smaller fans around the perimeter of a simple open plenum craft. Such a multiple arrangement however would impose other problems due to complexity, but then I reasoned, what if only four opposed fans were used ? Predictably a little of both phenomenon i.e. pressure drop and pressure rise would be obtained. In order to ascertain this, some purpose built rigs were set up using a styrene plenum box with not too deep vertical walls on the *outside* of which were fitted four electrically powered fans. In order to facilitate initial flow into the plenum the box was raised above the surface by half an inch or so, the rig in this instance being mass balanced via support lines and pulleys. Not only did this set up work, but it introduced me to my first recirculation fed VTOL vehicle. The function of this is best explained by the following analogy. A helicopter is hovering above a flat surface. The descending air has acquired energy from the rotor in exchange for the rotor lift. As with the foregoing open plenum, the descending column of air contacts the flat surface and is entrained through three hundred and sixty degrees in every direction. However this energy is *horizontally* lost to the surroundings, the magnitude of which can be experienced by any nearby spectators as standing in a gale force wind.

Now in this visual, but not very likely experiment, imagine four such helicopters hovering a significant distance apart. It follows that although much of the descending air flow from each has been lost to the surroundings as before, that in the middle of the four helicopters there is a considerable amount of air moving horizontally towards a common centre, whence upon interacting it is forced to be redirected vertically as an ascending single column of highly energetic air.

The next part of this unlikely experiment requires that we imagine the four hovering helicopters to be inter connected by a fixed cruciform on to which is mounted our original open upturned plenum dish, so strategically placed as to intercept this ascending air. This is now forced to split radially towards the helicopters whence it is interrupted by the vertical walls of the centrally situated dish, where once again it is redirected downwards as a hovercraft type peripheral sheet.

Our research revealed this 'reclaimed' recirculated air energy amounted to no less than forty per cent of the original thrust. I loosely termed the principle as 'recirculation lift' and it is that which I developed for my Hoverplane concept shown later on. Most importantly it will be apparent that the recirculation phenomenon requires that there is significant distance between the downward moving air streams and not least unlike conventional hovercraft the process also requires significant height in which to become manifest. This also will be evident in the hoverplane section. I also discovered that the effect could be maintained by six descending plumes of air, one at each corner of a rectangular planform and the other two interspersed at the centre. The four fans arrangement we commonly called a 'four poster' which interestingly is similar to the Harrier VTOL aircraft. In some instances two plumes can be encouraged to produce the recirculation effect. In this context, due to the fact that the four post arrangement of the Harrier was more suitable for the integration of an open plenum, that aircraft did in fact suffer a degree of negative lift due to the proximity of wings and interface.

Flight testing in time

In 1958 of the various disc type aircraft models I built embodying this integrated lift system, the definitive version was to be a six foot diameter R/C free flight model employing the 'four poster' principle, this format chosen in order to minimize the weight and which could be powered by two or more uncoupled engines. I was familiar with the necessary control/stabilizing systems as employed in rockets, including the Saunders Roe Black Knight, and despite the elegance of such systems, the miniaturization of these was practically impossible in those days. So I was well aware of the fact that I would have to rely on some tethered hovering flight finger tweaking. As I had given this a great deal of thought perhaps the following dream would have been of little significance. However, as will be seen later on in this section, no less than twenty six years later in 1981 the physical content involved certainly was.

In this dream the completed six foot diameter model was flying inside a building, in fact of all places right in the middle of our drawing office at Saunders Roe, where the average space between drawing boards and desks wouldn't have been large enough to have even set it down. It was hovering slightly lower than my head requiring me to stoop a little beneath it as I casually finger tweaked the stability controls. Now this model above me was perfectly steady in all axes, so why I was fiddling with it I don't know. Moreover there was no noise, nor oily exhaust fumes as would be expected. The following day I reflected on this not unpleasant dream and had put it down to wishful thinking. As shown in Chapter 9 the reader might like to share the significant spread out time effect implied by considering the sequel as it really turned out in 1981!

AERONAUTICS   APRIL 1958

IMPORTANT ANNOUNCEMENT
Are 'saucers' so silly?

Despite wild stories in certain newspapers, and despite the intentional building up of the 'flying saucer mystery' for purposes of sales promotion, saucer-shaped aircraft are no novelty. Circular and annular wings, and other forms of disk aircraft, have been a subject of research for half a century, and the final chapter of this long story has not yet been told.

Research into disk wings continues both here and abroad, some with government support, some entirely in the hands of private investigators.

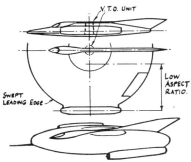

AERONAUTICS has now been able to acquire the full story of one such research programme which points the way towards new kinds of aircraft embodying structural as well as aerodynamic advances.

The illustrated report of this intriguing effort by a leading worker in this field of research will appear in next month's AERONAUTICS.

Saucerer. Co-axial, bi-convex to you, the unidentified flying objects with which Mr L G Cramp's latest inventions are associated, are in the news again. We hope to print an article setting forth Mr Cramp's unorthodox views in a future issue of AERONAUTICS. Mr Cramp is here seen outside Braxton Lodge, in the Isle of Wight, where he has been conducting his experiments

FLIGHT International,
28 March 1963

WORLD NEWS...

Russia's Discoplan II

The character of the "air-cushion glider" illustrated in our March 7 issue is partly clarified by a Soviet statement:—

"A circular-wing glider—*Discoplan II*—was successfully tested on an airfield near Moscow last autumn. The machine slowly taxied [presumably it has an engine—Ed] to the start of the runway and stopped. Then almost without any run-up it took off, resembling the notorious "flying saucer." The trials were conducted by a test crew led by M. V. Sukhanov.

"The circular wing has neither ribs nor longerons. The prototype for the wing was a bicycle wheel. The light-alloy edging around the perimeter of the "wheel" serves only to give a streamlined form to the tip. The glider is controlled by a rudder and ailerons. The wing diameter is 5m (16ft 5in), while the gross weight is 240kg (529lb). Thus, the specific load on the wing is the minimum—only 12kg/m² (2.46 lb/sq ft). *Discoplan II* was piloted by Vladimir Ivanov, a glider expert.

"One of the main features of the machine discovered during tests was the aerodynamic effect of the 'air cushion.' Owing to the proximity of the wing to the ground, and thanks to its circular form, the influence of the cushion on the landing and take-off characteristics proved most beneficial. At a height of 1.5-2m (59in-79in) *Discoplan II* automatically stabilized both in the lateral and in the transverse directions and could fly without the pilot touching the controls. In gliding down to land the pilot also felt how the discoplane was 'padded' by the air cushion. Thus, the landing was also of an automatic nature.

"Some experts believe that the advantages of the discoplane can be used to full advantage at high speeds in VTOL craft. The high manoeuvrability, combined with the anti-spinning properties and safety of take-off and landing, may prove exceedingly useful for training and other purposes.

The extraordinary similarity between the author's independent observations of merits for the disc wing is made clear here by comparing the above Flight International article in 1963 on the Russian Discoplane in which even the descriptive constructional terminology is duplicated as underlined in the author's article in Aeronautics 1958 page 84. Viz

The Bicycle Wheel!

The disk-type aircraft

Leonard G Cramp, A R Ae S, makes here a report on a disk-type vtol aircraft design which incorporates an inherently lightweight structure

Past and recent research into aircraft incorporating disk-type planforms has revealed that few advantages are to be gained, either structurally or aerodynamically, and for these reasons it has been rejected. However, the object of this article is to urge that the case be given further investigation, for it is felt that the disk-type aircraft does, in fact, possess certain inherent advantages, briefly outlined in this text, which have been overlooked in the past.

The following brief summary shows some of the advantages which are a natural consequence to an aircraft employing this type of wing.

(1) It has been found that a wing so constructed offers a vastly superior strength-weight ratio to more conventional types

(2) The design inherently advocates the employment of vertical take-off and descent

(3) Because of the extremely light structure plus v t o characteristics, the aircraft offers its crew greater chances of survival than those now accepted

(4) It has now been established that the disk-type aircraft offers the best compromise for re-entry into the earth's atmosphere

(5) In addition to the above, it is suggested that by rotating the disk wing, leading edge aerodynamic heating may be considerably delayed

(6) Due to the extremely light structure, the aircraft would have a greater payload capacity and/or greater range

(7) It is believed that this type of aircraft does not require a conventional fin, rudder and tailplane assembly, which again offers a considerable weight saving

(8) As practically the whole aircraft contributes to lift in forward flight, most of the dead weight of a conventional fuselage is largely eliminated ; as in a normal flying wing

(9) Being a v t o aircraft, a conventional type undercarriage is unnecessary, small castering type shock absorbers would suffice. This, together with a greatly reduced hydraulic system, also represents a considerable saving in weight

(10) Due to the total wing area being formed into a circle, obviating the need for fuselage and tailplanes, this type of aircraft would greatly facilitate stowage in restricted spaces, for instance, aircraft carrier hangar decks.

It is accepted that some of the above items could be true of any other ' flying wing ' type of aircraft, but the prime reason why they would be especially

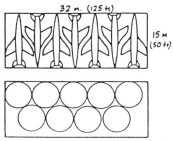

Figure —This shows how conventional aircraft and disk-type aircraft of greater area might be stowed in an equal amount of space

In the past, employment of a plano bi-convex wing would not have been considered, but with present-day supersonic aircraft it can be shown to have certain advantages.

The basic proposal of this incorporated design suggests that a plane circular wing be constructed in much the same fashion as an ordinary bicycle wheel, which, because all of its interconnecting members are under a stress loading, is immensely strong and resilient. It follows, to illustrate the principle simply, that a glider-type aircraft could be built comprising little else than an outer rim, central disk or cupola, connected by a series of turnbuckle tensioned wires, the whole then being covered with doped fabric. This structure employs neither ribs nor spars, but in fact is the stronger.

By way of example, a model aircraft with a wing span of 2·1 m (7 ft) and a chord of 30 cm (1 ft) having a wing area of 0·6 m² (7 ft²) has an all-up weight of 2·9 kg (6·5 lb), giving a wing loading of some 4·2 kg m² (14 oz ft²). An 0·9 m

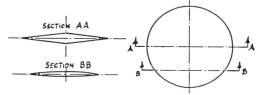

Figure —This shows the section of a circular-winged aeroplane according to the author's ideas. The section B-B is part of a parabola, and forms a bi-convex wing section

strength-weight ratio to more conventional types

applicable to this design is discussed below.

Normally when considering a wing planform the aircraft performance as a whole is a determining factor, and in the past the disk type or circular wing has received its share of consideration and, as previously stated, has been turned down on its merits.

This choice may have been justified from a purely low speed point of view, but speculation in terms of high Mach numbers brings to light new factors.

Figure —This depicts the wing section more graphically

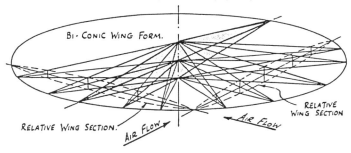

BI-CONIC WING FORM.

RELATIVE WING SECTION.   AIR FLOW   AIR FLOW   RELATIVE WING SECTION

Extracts from an article published in Aeronautics 1958 following a visit to the author's workshop by the late Oliver Stewart original commentator for the Farnborough Air Show, UK.

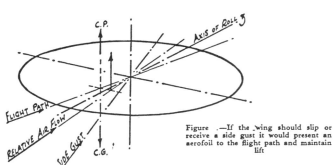

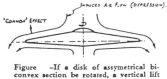

Figure .—If a disk of assymetrical bi-convex section be rotated, a vertical lift component due to the 'Coanda' effect would be generated

Figure .—If the wing should slip or receive a side gust it would present an aerofoil to the flight path and maintain lift

(3 ft) diameter circular wing having approximately 0·6 m² (7 ft²) of wing area was built on the above principle, and weighed some 0·425 kg (15 oz), giving a wing loading of about 0·59 kg/m² (/ft²). This model, dropped from a c derable height on to the rim, sus-t barely any damage. The weight saving in this case was no less than eighty-five per cent.

Figure 2 shows a cross section A-A through such a structure, represented by an upper and lower cone. It follows that section B-B is, in fact, part of a parabola and forms a bi-convex wing section. This holds good for approximately two-thirds the radius of the wing,

It is suggested that it would not be necessary to rotate the whole aircraft on change of course, reorientation of the centre portion may suffice. Should a disk of asymmetrical bi-convex section be rotated, a vertical lift component due to the 'Coanda' effect would be generated, Figure 5, the exact value of which could be determined by experiment. It is also suggested that a rotating disk may suffer less drag in a moving fluid than a stationary one, due to the local pressure rise at the 'trailing' edge, in Figure 6.

A limited amount of work has been done in this respect, without any satisfactory conclusions being reached. But the theory indicates that the combined

To this end, wing leading and trailing-edge flaps on more conventional type aircraft have been successfully exploited.

Figure 13 shows this to be in .effect an attempt to change a high fineness wing section to one with high-lift characteristics, which, of course, has its mechanical

(Turn to page 88 for Figures to 12 and to page 91 for continuation of text.)

Figure .—Varying sections of a circular-winged aeroplane are shown again here

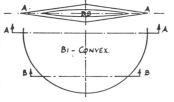

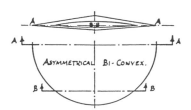

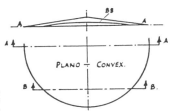

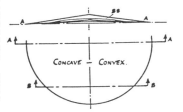

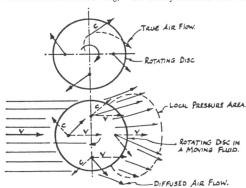

Figure .—A rotating disk might suffer less drag in a moving fluid than a stationary one, due to the local pressure rise at the trailing edge

gradually changing to section A-A. Figure 3 depicts this more graphically.

It must be stressed that this is a natural function of a plane conical shape, therefore no formed ribs are necessary ; the bi-convex shape being relative to airstream, which can be in *any* direction. A feature of this wing is the fact that it could even be rotating as stated, but because of its uniformity it presents an aerofoil to any flight path.

Due to this, should the aircraft slip or receive a side gust, it meets the wing normal, and lift is maintained—Figure 4. There are certain rolling characteristics associated with this, but these can be compensated.

effects of centrifugal and lineal flow induce a divergent and consequently decelerated air stream with an accompanying pressure rise. To some useful amount, this may offset normal drag.

Figures 11 and 12 depict some possible arrangements for v t o l, which are of a self-explanatory nature. However, in this respect there is another more attractive possibility—also an inherent function of the conic wing—which, it is felt, warrants further discussion.

As aircraft flying speeds have increased, one of the accompanying disadvantages has been the correspondingly increased landing and take-off speeds, which have in turn demanded longer runways.

limitations as indicated. Consequently, aircraft designers are looking more and more to the v t o l aircraft as the solution.

It is suggested that a variant of the conic wing aircraft might usefully contribute to that solution and is briefly outlined as follows :

Figure 14 shows a modified bi-conic wing having an annulus in the centre, which is formed by a radiused wall, section A-A. It follows that—as with section B-B in Figure 2—a development at the section making a tangent with the inner annulus, reveals a naturally formed high lift type aerofoil section of usable shape. This would seem an advantage, as it has been arrived at without the employment of complex and weighty mechanisms. It follows, therefore, that if an airstream is passed outwards over the wing, tracing a near tangential path to the annulus, a usable lift should be generated. Work has been done on this and the system works quite successfully, although complications are experienced due to the rather unusual divergent flow conditions over the wing, and the lack of data concerning it. It will be appreciated that between sections A-A and B-B, Figure 14, there are effective aerofoil sections of varying thickness/chord ratios which can be selected by a variable nozzle type cascade device such as that employed by the writer.

Unlike the comparatively thin trailing edge of the conventional wing, the disk wing leading-trailing edge is more rounded, and to some extent lift is augmented by the downward deflection due to 'Coanda' effect over this part of the wing as in Figure 5. It is possible that the effect might be even further enhanced by the provision of an annular flap situated at the periphery of the disk.

In the v t o l rôle the centre of lift would normally be situated at the centre of the wing which would demand a similarly placed c g position. With the aid of the variable guide vanes, however, an asymmetrically-placed centre of lift can be arranged, giving the aircraft a bias in any chosen direction. Work has been done in this respect, and the result suggests that at low speeds helicopter type manœuvrability may be possible.

Rotating a portion of the wing would offer an attractive alternative means of stabilising the vehicle, particularly when operating in rarefied atmosphere at high altitudes.

Airflow conditions from radial to longitudinal, such as would be experienced in transitional flight, may not involve insurmountable difficulties, as might be expected, but careful matching of the radial airstream velocity over the wing, and that due to the forward velocity of the aircraft will be necessary.

Figure 15 gives a diagrammatic idea of the general conception of hovering, transitional and directional control technique as at present visualised by the writer.

Only the simplest structure may be necessary in a high Mach number aircraft of this type, the two conical sections perhaps being fabricated from sheet

On this page the text from page 87 is continued. Figures   to   are on page 8S.

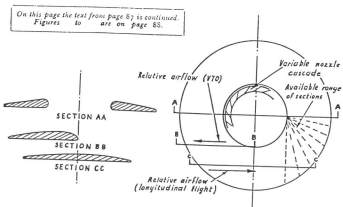

Figure 14—This shows various sections through a plano-convex annular disk wing

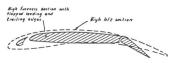

Figure   —The author considers that while transition from one wing function to another is reasonably easy, the flapped wing is at best only a poor compromise

steel, having a surrounding built-up rim leading-trailing edge. Such a structure would permit a wide range of internal layout without necessitating a comparable amount of structural alteration.

Perhaps it will be of some interest to point out that present research in magneto-hydrodynamics suggests the shape of wings to come. There can be little doubt that experience gained now with an aerodynamic counter-part will pay dividends later.

Footnote : Protective rights have been filed on the conic wing.

Figure   —This gives a diagrammatic idea of the general principles of operation of the disk-type wing machine

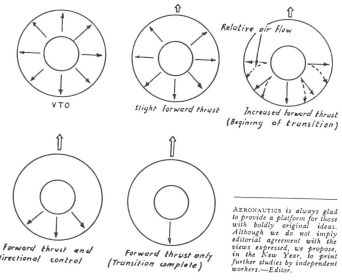

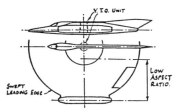

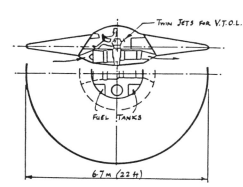

Figure   —This is one kind of circular-winged aeroplane that might be built for military or other purposes

Figure   — Several arrangements for a vertical take-off aeroplane with a circular wing are under consideration, and some of these arrangements are shown in these diagrams

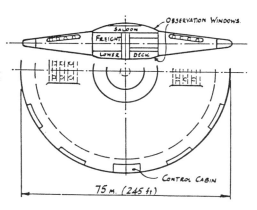

Figure   —This aircraft, similar to that depicted above, might carry freight in its central section, while having also an upper saloon and a usable lower deck. Observation windows would be fitted in the top centre

Figure   —Yet another type is shown here. It could be powered by twin turbojet units with a deflection arrangement for vertical take-off

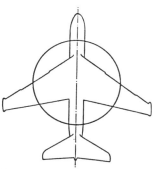

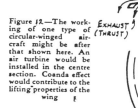

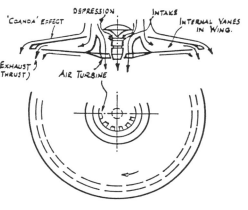

Figure 12—The working of one type of circular-winged aircraft might be after that shown here. An air turbine would be installed in the centre section. Coanda effect would contribute to the lifting properties of the wing

Figure   —A similar area could be encompassed in a circular wing as in a conventionally winged aeroplane, as illustrated here

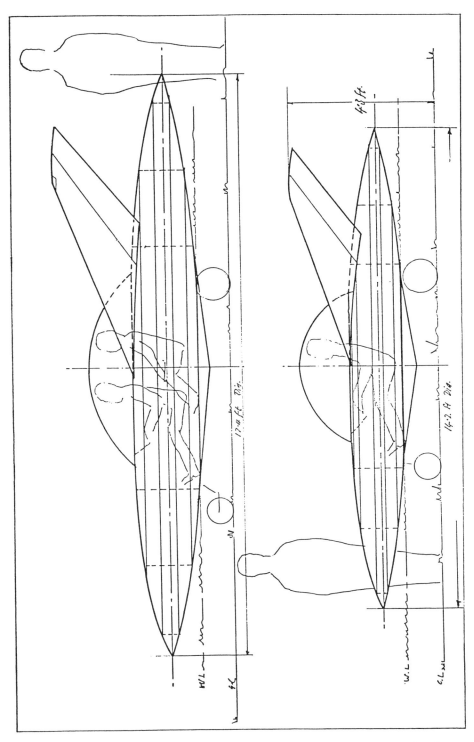

Two and four seat VTOL Rotorcraft. Control in all three axes is monitored with
automatic rate sensor inputs to variable incidence rotor blades.

Above and right. An investigation into combined disc wing and integrated powered lift rotor utilising surface effect. Below. 6ft diameter part structure illustrates simple construction of the disc wing. Below right. Sectioned model made by author's son Gary shows three main inherent aerofoils.

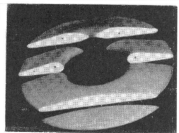

Above, VTOL model in flight.

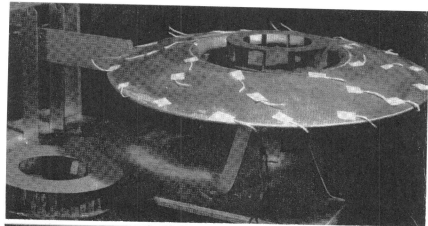

Above. One of the author's variable cascade disc wings undergoing lift evaluation tests.

One of four integrated rudder and elevator units for torque, pitch, roll and heave for a disc wing in which ground effect was used.

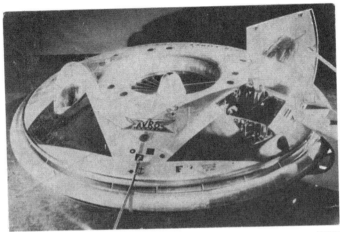

VTOL from Canada

ALTHOUGH the Canadian Avrocar has been under development and test for some time, very little information or photographs have been released. Designated Model 1 by Avro Canada and the VZ-9V by the U.S. military, the Avrocar was designed to explore the potential of a new approach to VTOL. It is strictly a research aircraft and is at present under the sponsorship of the U.S.A.F. and U.S. Army.

The disc-shaped Avrocar uses the air-cushion principle for VTOL but upon reaching a certain translational velocity, it functions as a conventional aircraft supported by aerodynamic forces generated by its forward motion. It has three Continental J-69 turbojets which provide the power for a turbine which in turn operates the centrally located, large-diameter fan. This fan supplies air for the annular jet around the rim to generate the air cushion close to the ground. Unlike other ground cushion vehicles, however, the Avrocar is also capable of high forward speeds at high altitudes.

One of the Avrocars has been tested in the full-scale wind tunnel at Ames Laboratory, Moffett Field, Calif.

From an article published in The Aeroplane August 12th 1960, even a brief examination of the photo and accompanying text reveals an astonishing identity with Fig.12 and text in the author's Aeronautics article of December 1958. It should be pointed out that the publisher received this material in 1957, while the original investigative work was already underway in 1953.

1960 - 1993 Proposed high speed track secured aircraft (TSA) as a cost effective & environmentally directed alternative to cross country public transport.

Basically a monorail system with an important difference, in that the vehicle takes advantage of the efficiency and merits of air travel, while it remains `tethered' to the surface,

As already discussed, it has long been known that aircraft skimming close to the surface can be more efficient due to reduced induced drag effects brought about by interaction between the craft and the surface, in other words ground effect. Unfortunately, like birds, freeborne aircraft can only take advantage of this phenomenon during take off and landing. The author proposed that a hybrid vehicle, such as the TSA should be able to employ this advantage throughout its entire operating range.

During the 1960s I was involved with the proposed, much talked about, tracked hovercraft train, which was to be supported on a monorail system by multiple high pressure air pads, each formed much like miniature hovercraft. The vehicle was to be conventionally propelled by fans and/or the well known Laithwait electromagnetic linear motor system.

At that time some designers were concerned about the considerable aerodynamic lift to which the streamlined vehicle would be subject, which might tend to lift off the air bearing pads, thereby interfering with the cushion pressure efficiency.

It occurred to me that rather than try to eliminate this problem it might be profitable to consider doing an about face and actually using it to advantage. This could be achieved by designing the vehicle as a highly refined aerodynamic lifting body which at speed and low angle of attack would generate lift equal to most of the weight of the machine, leaving only a measured proportion to be taken by a monorail track via air lubricated pads. This is of course the antithesis of the foregoing tracked hovercraft in which the *entire* weight of the vehicle is sustained by the track.

It is further proposed that any track variation will be monitored and computer fed to the TSA conventional on board flying control system. As would be vertical and lateral acceleration, including pitch, roll and yaw changes. Thus the machine will be `flying' continually on automatic pilot. The system will offset centrifugal loads due to track curvature by banking the machine in exactly the same manner as if it were free airborne as similar to the British Rail APT (Advanced Passenger Tilt Train) now being considered. It is assumed the track mounting system would be so designed that in no circumstances - other than intentional - can it become detached. Thus passengers will be able to enjoy the pleasure of air travel without leaving the ground, which would be good news to all those who are nervous about flying.

Cascade wing and
surface effect

In addition to the lifting body configuration, the TSA would embody the cascade wing lift augmentation as previously shown with the Skimmer craft. This, together with hovercraft type cushion formed beneath the vehicle by deflected propulsion air, would produce static lift to initiate lift off prior to forward tethered flight mode. Moreover in so far as some of the rail system engineering requirements (vehicle routes and transit times et.) envisaged for the TSA are similar to those for the earlier proposed tracked hovercraft, it is useful to review appropriate data (shown in the accompanying diagrams).

At that time it was generally taken for granted that on some routes the hovercraft tracked system would have to negotiate large stretches of water and other obstructions, e.g. existing roads etc., for which overhead stretches of track would have to be built. The latter would also apply to the TSA where roads and narrow rivers are concerned. But over larger stretches of water, such craft would be capable of operating off the track as free airborne skimmers (similar to that shown earlier) then automatically reconnected to the monorail for the remainder of the journey. Such a vehicle would obviate the construction of a tunnel or bridge between the mainland and the Isle of Wight for instance, permitting transit times across the Island and Solent up to London in little more than an hour for the entire trip, including the stops.

Partly under the auspices of a large Aerospace company, the TSA concept is currently being investigated with the aid of scale models, by a group of engineering students at a local High School.

TSA 100 in free airborne channel crossing mode.

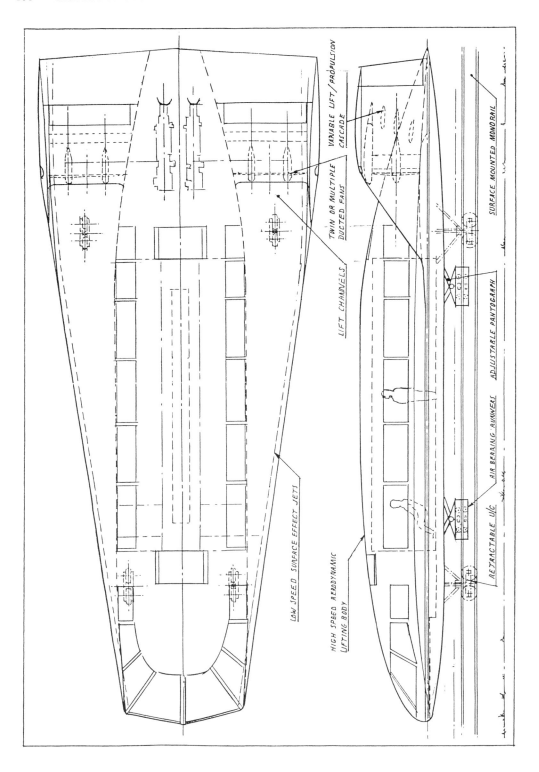

G.A. TSA 100 high speed aerodynamic lifting body monorail vehicle.

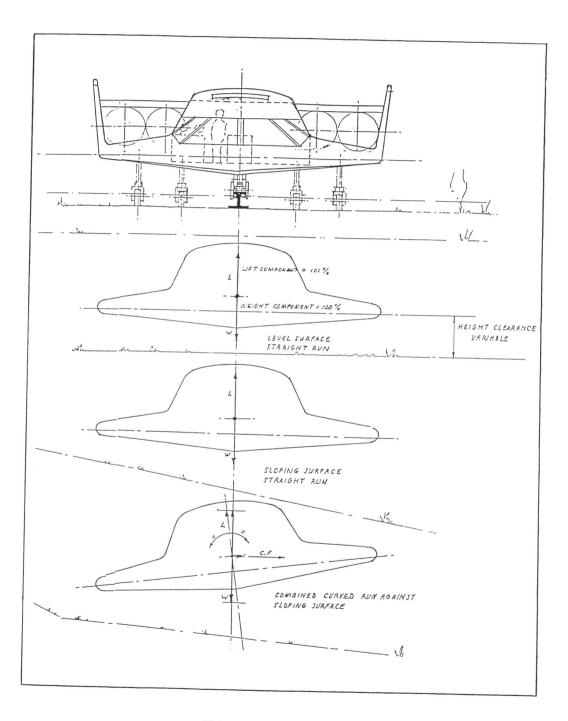

TSA terrain operating geometry

*Some possible 250 m.p.h. tracked-**Hovercraft** trunk routes from London, and the effective size of Gt. Britain so equipped. Journey times would be: Edinburgh 1hr 47min: Manchester 1hr: Birmingham 33min: and Bristol 37min.*

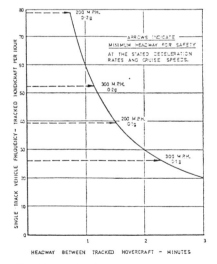

The relationship between track capacity, safe headway and vehicle characteristics.

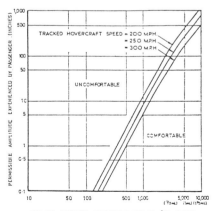

*Limiting vibration levels for passenger comfort at high tracked-**Hovercraft** speeds.*

Proposed tracked hovercraft data based on an article in Flight International August 1965.

TSA test model showing twin lift channels, fans and open cascade.

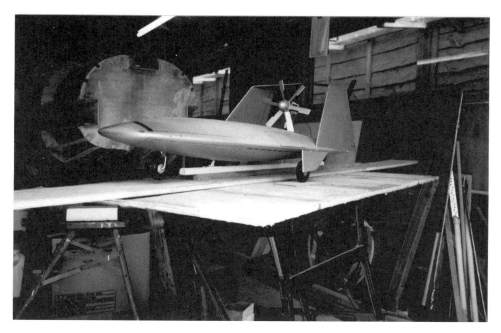

TSA lifting body test model grounded astride monorail track.

*TSA lifting body test model in static tethered flight mode
above monotrail track.*

8
Hoverplane theory 1ˢᵗ.Generation

1953 - 1992. This investigation included both ends of the size spectrum, the first manned test vehicle having flown in 1984. ATF 39 years.

The following is a compilation of extracts from various studies of the author's 'Hoverplane' concept, involving such organizations and institutions as The Cranfield Institute of Technology, The Wolfson Unit of Southampton University, the Department of Industry, BP, ICI, UK and various overseas interests including the Canadian Coastguard R&D Department, Aeronca and several industrial concerns from other parts of the world.

Initially, the Hoverplane was another attempt by the author to find a simple, fairly low-tech solution to some of the problems inherent in all hovercraft or surface effect vehicles, not least instability at increased height above the supporting surface. As previously stated, I was well acquainted with the problems way back in the 1950s when my son Gary suggested the flexible tape solution for the little 'Orbit' model. What was obviously required was a vehicle which could take advantage of the heavier load carrying capability of the surface effect principle, but which could also remain stable in roll and pitch at height, leaving it free in heave to negotiate rugged terrain surfaces. The Hoverplane is one such solution, all of which was derived from those first lengthy early excursions with models. Lengthy excursions over years, which could never have been meaningfully entertained by industry, universities or the like. It could only have been successfully attempted in the early stages of development by individuals like the author, in a private capacity. Indeed it was for such very curtailing reasons that Hovercraft, despite their unquestionable usefulness, became relegated by the British Hovercraft Corp., and others, to little more than flexible underbellied boats with special beaching capabilities. This contributed towards the decision for me to leave my job in the projects department at BHC in 1971, in order to continue other works shown in this book, including investigation into the 'up and over hedges' possibility of surface effect vehicles, as I first visualized them as a youngster and more definitively in 1943.

I cherish the memory of the occasion when I first surrendered to the almost childishly simple solution - as it seemed at the time - of rigidly securing a party balloon containing helium gas to a small hovercraft model we had built for pitch and roll testing. Due to the effectively reduced weight the model rose higher over the workbench, but this time not only was it perfectly stable, it was now impossible to destabilize.

By today's full size application standards it is now purely academic, but the next thing we saw - and to this day never tire of displaying - was the perfectly balanced model when arriving at the edge of the work bench, instead of toppling over, as would normally be the case, merely slipped gracefully off, descended to the workshop floor and literally bounced on a cushion of air without physical contact, and after a couple of these gentle 'pogo' type mini bounces and to the delight of us all, continued merrily away across the workshop floor! Fascinated, I realized that sometime somewhere I had viewed this very thing before. Eventually time went reeling back to that day when as a schoolboy I had visualized this very thing in that empty class room many years ago (see Chapter 1)

Another helping hand

Inspired by this 'pogo' hovercraft I built a modified rig comprising a small ducted fan with anti torque blades supported by four splayed legs surmounted by a helium filled toy balloon. When this was hovering out of surface effect it could literally be turned horizontally and let go, whereupon it would merely oscillate, pendulum style, several times before resuming the static hover mode. Thus it conformed exactly with predicted theory i.e. applied torque and couple laws.

Then on one notable occasion my youngest son David was present at one of these tests and on the point of leaving the workshop he looked at me and said " Dad, what would happen if the balloon contained only air, wouldn't it have the same effect ?" To which I hastily assured him that as the lift of the helium gas provided the correcting torque, the thing wouldn't work leaving the ducted fan completely unstable. David had acknowledged this hesitantly with a not very convinced sounding " I suppose so " and left the building. For a moment I watched him returning to the house, half wishing for his sake that he would have been right, but of course, he couldn't be.....could he?

Then looking at the contraption on the bench for a brief instant I saw what David had seen. I didn't believe it, but there was only one way to find out, so casting all reason aside I removed the helium filled balloon and replaced it with an air filled one. Then I had switched on the power supply and gradually increased the speed of the fan. Not only did the unit rise off the bench as before, but unbelievably when it reached the hover mode and I gingerly tilted it, hey presto David's intuition was right, for the unit stabilized immediately, other than a slight increase in lift, there was absolutely no difference. Today it would have taken only a few seconds of computer time to give the explanation, but at that time it took me several weeks to finally establishment it.

Several years later when the hoverplane concept was being discussed between our publicity manager and aerodynamicists at Bristol University, they had also challenged that stabilizing torque could be maintained with an air filled cell, that is until one of them interrupted their meeting to phone me and I was able to offer him my time worn explanatory analogy. Whereupon he thanked me and returned to the meeting. The following day the publicity manager told me that the young aerodynamicist had returned to the meeting, sat down and looking at the others said "He's damned right you know". I would have preferred that my son David had been given that acknowledgement.

During the ensuing time several sponsors were contacted which resulted in some limited funding. On average I worked fifteen hours a day, every day of the week at an average hourly rate of under £1! I had to accept that from by budget in order to pay my son Gary a basic minimum rate, therefore we were both on the breadline. Over the years on no less than three separate occasions, my bank overdraft and mortgage required putting our house on the property market. I usually managed to stall off this situation a little by overpricing the real value, for it would have spelt disaster for me to have lost my workshop in the country, where we could avoid annoying neighbours with our extended nightime labours.

Despite the hardships we often had great fun bench testing and test flying the models. During the winter it was common to work with top coats and gloves in order to keep warm. It costs quite a lot to heat a workshop of over two thousand four hundred square feet! Necessary test equipment the sponsors couldn't afford was designed and built by Gary from redundant scrap components purchased from such places as the Plessey Radar Works and scrap merchants. On the market one `multi-variac' power supply unit alone which would have cost several weeks salary, Gary freely built in his own spare time for the grand sum of £30! To this day I am sure the sponsors would still be unaware of these mini miracles. But as I have intimated elsewhere, the world of innovation can at best be quite indifferent and at worst damned hard.

However during this time we were successful in getting the interest of the engineers at BP's Research and Development establishment at Sunbury, which although not improving our financial situation, offered much encouragement to us all. The following is a compilation of extracts from reports required by them and others.

A special note
to modellers

It is fair to point out that as a very exacting modeller the author is sensitive about any lack of aesthetics with some of the models shown in this book, particularly in the case of the Hoverplane. This is primarily due to the fact that working with a tight budget leaves little room for refinement. Therefore much of which is shown here are little more than decorated test rigs. This tendency is further exacerbated by the out of proportion, ungainly inflated software, functionally adequate, but not very pleasing to the eye.

On the credit side however it should be pointed out the Hoverplane principle allows VTOL model beginners to fly and hover hands off *first time,* and as shown in some instances without helium gas! In other words, such craft are as easy to operate, as model helicopters are difficult, I know, for this requirement played no small part in my original motivations.

On the debit side, the necessary oversized inflated lift cell can set a limitation on operation in windy conditions. But a fairly generous fin area helps to negate this problem by keeping the nose into wind on landings and take offs. Incidentally, this problem is much reduced in full size hoverplanes and it is prudent to emphasize to critics, a generous sized air or helium filled delta wing represents a considerable *inertial* resistance to wind gusts while in hover, which can vary from a few pounds in models to hundreds of pounds in full size applications.

A case of anti-gravity

This introduction to the hoverplane would be incomplete without mention of the fact that although in technological terms the principle of hydrostatic lift as opposed to so-called anti-gravity might seemingly be light years apart, as I have stated elsewhere there is in fact a very close identity between the two. For a balloon merely obtains lift, or displacement, due to a direct reaction to the downward effect of gravity. Indeed I went to some length to amplify this in my book The Cosmic Matrix. So it is in no way incongruous that my research in the phenomenon of gravity should have included this Archimedian phenomenon, for as a student, to me it was one of the most logical first steps to take.

The Hoverplane Application

The Hoverplane is a revolutionary aerial transport concept which combines overload carrying advantages of surface effect vehicles, with vertical take off and landing and a unique stabilizing principle. Its application ranges from small ultralight sports craft to heavy load carrying rough terrain transports of many tons, as largely portrayed in this report.

Historical note

It was erroneous to designate Air Cushion Vehicles (ACVs) as hovercraft, for the simple reason that such craft do not hover helicopter fashion, as an ACV operating in surface effect above weighing scales will quickly reveal. However it is anticipated that the label Hoverplane will be descriptively enough.

In 1956 when the hovercraft first became public, work on the *Hoverplane* was already under way and early designers had been quick to appreciate the advantages of a phenomenon called augmentation lift, that ACVs of significant power loading can inherit. But they just as quickly abandoned the advantage of the principle due to the narrow area of development they had persued. Fortunately work on the hoverplane did not cease and it will become evident later in this study that the lift augmentation phenomenon can really pay dividends when related to ACVs of higher power loading *in vertical flight*. Any ambitions early designers had in this area were also soon abandoned, partly due to the formidable inherent stabilizing problems. With the hoverplane this problem has been successfully and surprisingly simply overcome, as an examination of the following material will reveal.

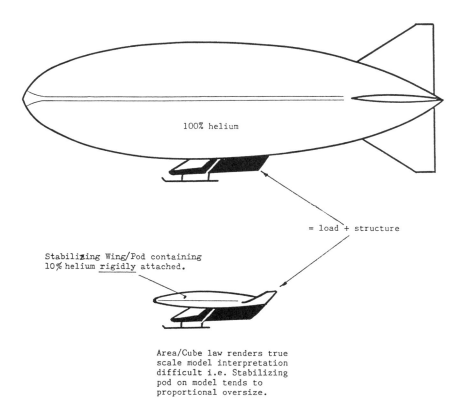

That the Hoverplane is not an airship is illustrated here more graphically in terms of relative volume required for a given load. Note, in some cases the use of helium lighter than air gas is optional. The stabilizing wing pod will still function as such if the contained volume of air is sufficient. However 10 - 20% of helium or more, usefully reduces the engine h.p. required for VTO.

A new versatile concept

At first glance this new innovation might understandably be interpreted as a composite between a conventional Air Cushion Vehicle and an Airship, supplemented by helicopter rotors similar to the Goodyear and Heli - Stat projects. But it is important to emphasize that such interpretations would be misleading, for the Hoverplane concept incorporates radically different approaches which enable the machine to be so versatile.

What the Hoverplane is not

Neither is the Hoverplane simply a hovercraft or `wing in ground effect' vehicle which has been made to fly, for in their present dynamic form most ACVs have little more chance of stable vertical flight than does a house brick, while the wing in ground effect machine cannot hover. In forward motion a degree of this type of lift does occur with the Hoverplane, as it does in fact with most aircraft flying close to the interface. In a word it is an inbuilt trade off advantage rather than specific design.

True hovercraft

In so far that this new innovation is capable of operating both near to the interface (ground or water) in surface effect and hovering helicopter style out of surface effect, it has more meaningful claim to the title 'Hovercraft' than any present ACV, for that is exactly what this new vehicle is. It is neither an aeroplane, helicopter or ACV but a true hovercraft.

The case for the Hoverplane, what it is & what it will do

The fact that designers have continuously considered combinations of airships and helicopters as heavy load lifters, in no better way illustrates forcibly the usefulness of the Hoverplane, for it is able to exploit the merits of both these alternatives, while not being penalized by the sheer size of the one and prohibitive cost and size limitations of the other. Additionally the Hoverplane can employ the unquestionable advantage of close surface effect lift to carry even greater loads, which clearly airships and helicopters cannot.

Variable mode and load capacity

As with all surface effect machines when lifting off with a given load, the Hoverplane will require increasing power with increasing air gap, (separation distance between underbody and interface). But supplemented by helium lift, the effective power loading of the Hoverplane enables it to climb above the plateau of useful surface effect to free flight (VTOL). Thus the vehicle can take advantage of very efficient close surface effect loading (around 60 lbs.per h.p.) and where necessary hover and VTOL loadings around 9 - 12 lbs per h.p.

Collective lift components

In operation the Hoverplane can be described as an amphibious hybrid vehicle which derives its total lift from three separate functions each of which is complimentary to the other. They are:-

1.   Dynamic thrust lift.

2. Hydrostatic lift.

3. Surface effect lift.

The proportions of all three being variable according to operational requirements.

Dynamic thrust lift

Unlike most ACVs, the Hoverplane obtains a significant proportion of its lift in static hover and VTOL with conventional axial flow fans or propellers, which are far cheaper than helicopter rotors and much lighter than hovercraft centrifugal fans.

Different surface effect

The surface effect technique employed in the Hoverplane differs significantly from that employed in modern ACVs, lift sequence being as follows.

Lift off sequence. Stage 1 & Stage 2. Close static cushion.

Propeller lift air in the form of two or more separate columns is strategically situated *outside* the perimeter of the machine and high above the interface. This automatically eliminates any tendency for blade stalling in overload and boggy surface situations suffered by ACVs. Thus the propellers working unrestricted by back pressure are able to produce maximum static thrust which effectively reduces the weight of the vehicle and initiates separation from the interface, Stage 1. This is followed by a consequent static cushion type pressure rise between the craft's underbody (plenum) and the interface which completes this part of the vertical take off sequence, Stage 2. Lift off sequence, Stage 3. Deep surface effect (from static cushion) to Stage 4 VTOL. For a given engine power input, due to the initial increased lift (lift augmentation), the machine continues to rise until the `static' cushion is replaced by controlled recirculation of the lift air between the interface and plenum space, causing an enhanced momentum exchange which can be in the same order or even greater than the `static' type cushion lift. But with one significant and most important difference i.e. this effect can occure at very large air gaps, up to the length of the craft above the interface, Stage 3. Accordingly the Hoverplane is able to operate in deep surface effect which permits it to take large obstacles and deep crevices in its stride without the introduction of deformable flexible skirts.   The manoeuvre can then be extended to Stage 4 (VTOL) when required.

Hydrostatic lift and variable geometric metacentre

The hydrostatic lift component is provided by the extremely tough inflatable wing/pillow containing an inert, lighter than air gas (helium) situated at a controlled height above the vehicle's centre of gravity (metacentre). The displacement or lift due to this is nominally around 18 - 25% of the gross weight. Thus power input for lift is proportionally reduced. Provision can also be made for a variable metacentric height system.

Inexpensive inbuilt stabilizing system

It cannot be over emphasized that any surface effect and VTOL advantages of the Hoverplane would normally be of academic interest due to the tendency for all surface effect machines and indeed all hovering machines, to be inherently unstable at significant hover heights, but for the simple innovation of the stabilizing pillow/wing, which obviates the necessity for otherwise complicated and sometimes costly dynamic alternatives.

Variable stability pillow

It is anticipated that the additional useful load lifting variable afforded by the stability pillow will be kept to a governed proportion when the machine is working in surface effect, for oversize of the pillow will cause induced adverse interface suction effects.

Note; in order to function as such it is imperative for the stabilizing/lift pillow to be rigidly mounted to the vehicle's hard structure as portrayed in the accompanying photographs and drawings.

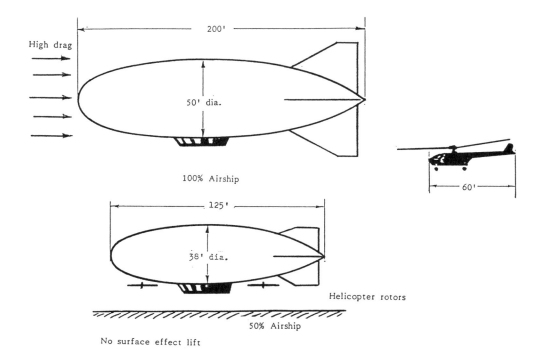

High drag

200'

50' dia.

100% Airship

60'

125'

38' dia.

Helicopter rotors

50% Airship

No surface effect lift

Airships size/lift relationship for 6 tonne load capacity equivalence.

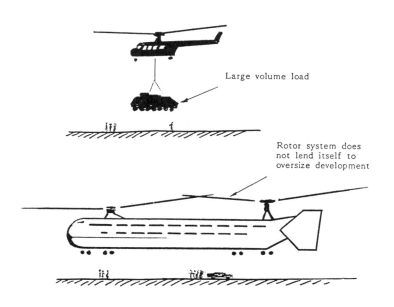

Large volume load

Rotor system does
not lend itself to
oversize development

Helicopter load/size restrictions.

Hovercraft terrain limitations.

Section A–A

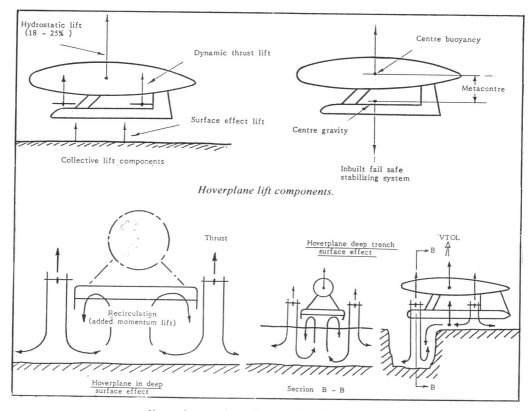

Hoverplane lift components.

Hoverplane surface effect and lift off sequence.

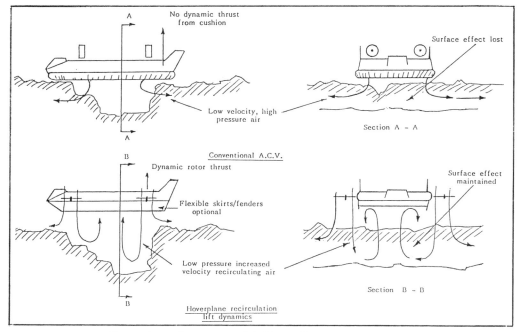

Comparison of conventional A.C.V. and hoverplane lift systems.

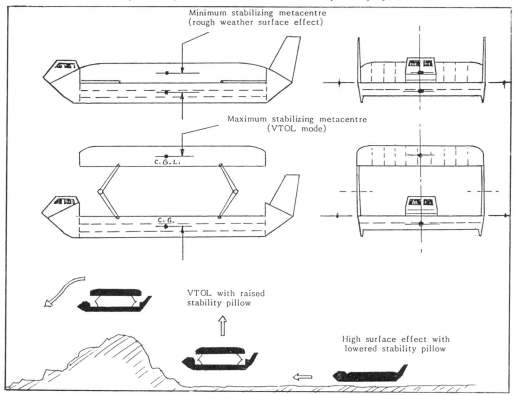

Variable geometry lift/stability pillow.

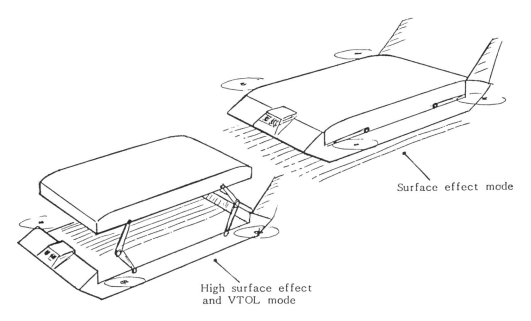

Schematic arrangement of variable geometry stability pillow.

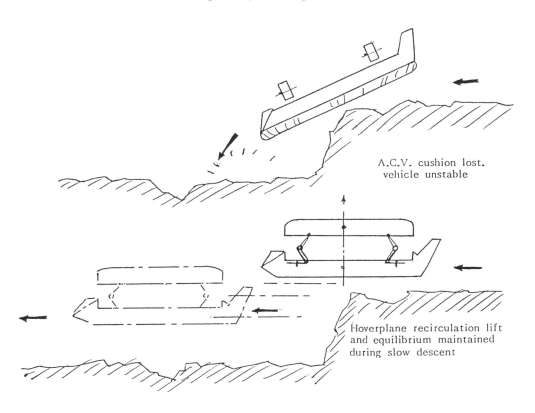

Hoverplane equilibrium maintained during slow descent.

1954. The development of the first deltoid version of the hoverplane.

This small runabout version of the Hoverplane was fitted with two swiveling ducted propellers providing vectored thrust for vertical take off and landing (VTOL), supplemented by the Cockerell hovercraft type peripheral jet for surface effect lift.   A stability/lift pillow containing helium gas at approximately 25% of the all up weight, also acts as a deltoid lifting body type wing of generous proportions, this has been omitted in the following photographs.

When Desmond Norman of Britten Norman Aircraft and a prominent member of the CAA first saw the static operational model of this craft they stated it was among the most fascinating ideas they had ever seen.

*One of the first test rigs of
the hoverplane.*

Test fuselage.

Swivelling ducts in VTOL mode.

Forward flight mode.

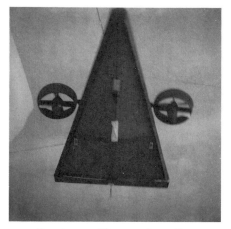

*Showing auxilliary surface effect
peripheral jet.*

*These photos, representing several models,
typifies the natural development of the
deltoid Hoverplane configuration, in which
toy balloons, converted bin liners and
eventual fully lofted and manufactured lift
cells took their place.*

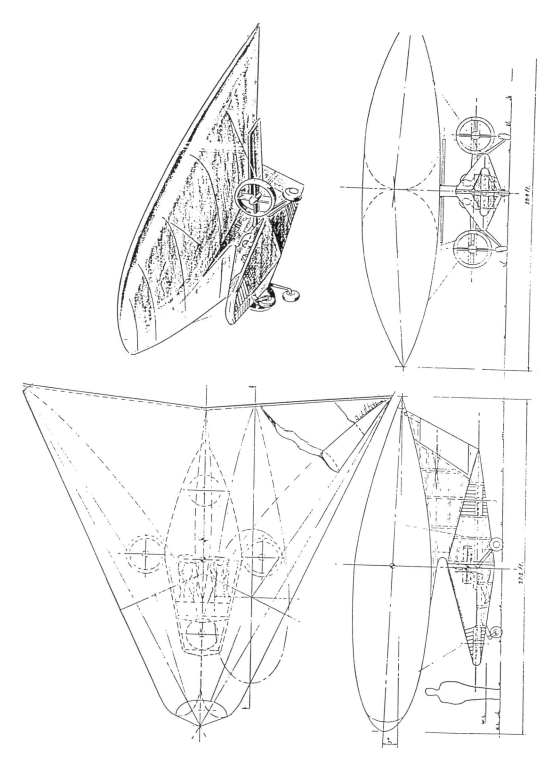

General arrangement of deltoid hoverplane 1954.

1956. Humming Bird four seat variant of the hoverplane.

In this version of the Hoverplane helium hydrostatic stability lift was provided in two streamlined wing tip pods supported by short conventional winglets in which were housed two swivelling ducted fans for vectored thrust. The metacentric height of the pods could be varied by the simple expedient of altering the dihedral angle of the supporting winglets.

Unfortunately due to change of interests abroad this project was shelved at the radio controlled model stage.

Head on view of Humming Bird hoverplane showing wing tip lift cells and swivelling ducted lift/ propulsion fans.

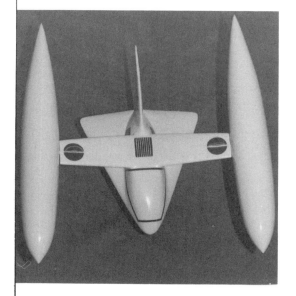

Plan view showing the low drag configuration and deltoid surface effect fuselage.

Three quarter view of Humming Bird shows one of several experimental generous area fins.

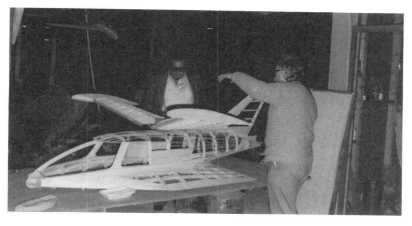

In this arrangement the requisite metacentric height between the centre of hydrostatic lift and the centre of gravity of the vehicle was achieved by a high mounted mainplane with a generous dyhedral angle. This, together with a forward swept wing planform, also achieved the fore and aft design location for the drag/lift components of the wing tip lift cells..

Quarter scale R/C model of 'Humming Bird' with main structure nearing completion.

1953. The Torus version of the hoverplane. A R/C model was successfully demonstrated at BP's R/D establishment at Sunbury UK 1978.

It should be noted that due to the area cube law. The inflatable lift cell/wing of all hoverplane models and in particular the `Torus', are penalized by a necessarily over sized representation.

The following variant of the hoverplane principle takes advantage of the Lippisch type lift generated above the large bell mouth intake formed by the circular torus lift pillow and the fan placed centrally within it. In this simple, inexpensive and fail safe arrangement, surface effect is inherently formed by the radially flowing air flow beneath the cabin. The vehicle is also capable of descending parawing fashion from hover or forward flight as are other variants of the hoverplane. Forward flight speed is somewhat penalized by the Torus, but its useful roles as a cheap runabout, police traffic surveillance vehicle etc., are obvious. Furthermore it has been shown that it can be flown after only one hour's tuition, microlight style, and it will cost a fraction of that of the helicopter. As with the other variants of the hoverplane, `Torus' is stable in free fall descent when the air flow is reversed.

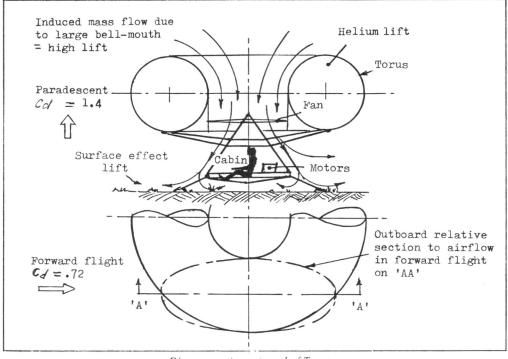

Induced mass flow due
to large bell-mouth
= high lift

Helium lift

Torus

Paradescent
$C_d = 1.4$

Fan

Surface effect
lift

Cabin

Motors

Outboard relative
section to airflow
in forward flight
on 'AA'

Forward flight
$C_d = .72$

'A'   'A'

Diagrammatic portrayal of Torus.

Torus in surface effect and VTOL/hover mode.

*Cabin and torus coupling support arms
of electrically powered model Torus.*

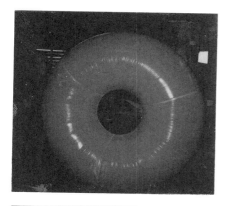

Eight ft. dia.
Torus before
assembly.

Cabin and power unit
of quarter scale R/C
'Torus' model as dem-
onstrated at B.P.
Sunbury R/D estab-
lishment in 1978.

Engine installation bay
with fuel tank and
nacelle removed. One
of several anti-torque/
yaw control blades is
visible.

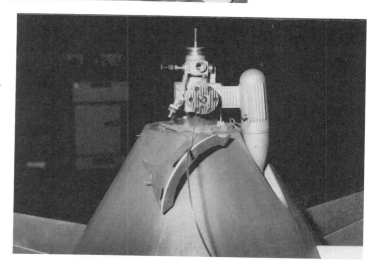

1982. Skylark.Light VTOL aircraft version of the hoverplane.

This format of the Hoverplane was presented to the technical staff and board at Cranfield College of Aeronautics in 1982. As with other versions, this craft was so forgivingly docile it was capable of paralanding vertically with both power and `hands off'. The third scale model was powered by a single engine with a swivelling propeller to provide vectored thrust. The lack of elegance of this research model was more than matched by its extremely impressive free flight performance. Full scale definitive design clean up would have improved this performance and ascetics considerably.

The transition time from wing lift in forward flight to vectored thrust lift at zero forward speed had been estimated to be around three or four seconds. However due to the incurred momentum drag generated by the ingested fan air, the transition occurred in just one and a half seconds. In other words we had found a very effective air brake!

It is estimated that the full size two seat version would have a top speed of around 150 mph while its selling price would have been less than one quarter of a comparable helicopter.

Superimposed precognitive dreams

In 1982 while in the process of building this model the author experienced a dream of the kind described in Chapter 1. In the dream my son and I were standing in the eleven acre field in which we conducted our model tests, Gary was flying a model. Bearing in mind what has already been said about the difficulties of underscaling the Hoverplane principle, all our models were consequently fairly large. Also it should be emphasized at that time all the lift cells were cigar shaped.

In the dream Gary had suddenly lost control of the model and it flew into some bordering trees. Immediately I was trying to retrieve it by parting the bushes and spotted a bright yellow object, it was delta shaped and about one foot long. In the dream the sudden transition from handling inflatables of several yards in length to this unexpected miniature version gave me quite a start, exacerbated by its extreme lightness.

Several weeks later we were trying to improve the aerodynamic lift for the Skylark, however with the manufacturing means available to us a true delta at one third scale would be almost impossible. Then I suddenly had the idea to compromise by joining two lop sided cigar shaped cells, which although still rather crude might give a better performance. Even so `Tuftain' material from which we made the cells was very expensive, tending to curtail extravagant full size trials, but this simple idea would at least permit an under sized aerodynamic test.Within a few hours I had made such a small `delta', inflated and ballasted it and chuck glided it across the workshop. It flew beautifully, so I hurried over and picked it up and immediately went cold. I had had exactly that tactile experience several weeks before! True the local environment was different, but the physical association was identical, in the dream it had merely been a small yellow delta wing. In reality it had become a small yellow inflated truncated delta formed by two simple lop sided cigars.

A forty year sequel to
a precognitive dream

As will be seen from the still photographs the idea worked sufficiently well enough to validate this simple Skylark version of the hoverplane. Among other things video footage shows it paradescending with a dead motor and hands off. However of no less importance to this book is the fact that because it was so easy to fly, one day I took control of the model, this, perhaps foolishly, because at that time my vision was seriously impaired by cataract. Consequently having successfully and no less joyously, flown the model in VTO and hovering modes, I took it through transition to forward flight at about ninety feet altitude, whereupon I was immediately dazzled by the setting sun and promptly `landed' Skylark in the very top of a seventy foot high pine tree! We could not afford to lose the model and the sponsors were due to arrive in a day or two. We had no equipment nor any means to reach the model other than by climbing the tree and the branches at that height would be very slender, not least I still suffered from a fear of heights. However despite my protests Gary had already began to climb the tree. With bated breath and fervent prayer I watched as he eventually reached the top, and how he managed to clamber out and release the model I shall never know, yet he did, and with a flash of *brilliant yellow from the setting sun* reflecting on the large Tuftain wing, Skylark came parascending down to me. I watched Gary's safe descent and between my words of thanks as we were assessing any damage incurred Gary said "It's a cracking view from up there Dad".....It would be several years before, from somewhere, `the penny dropped'.

Over the ensuing years I had on occasion remembered that dreamworld stranger with the golden red hair in 1942 I discussed in Chapter 1, who had so bravely rescued Gary,then our little son of only a few months old, until extraordinary as it may seem, I realized I had no difficulty in identifying the stranger with Gary as he would become forty years later.

Fact. Dream in 1942
1.The stranger had golden red hair.
2.The stranger safely retrieved my child from a perilous height bathed in sunset light.
3.The stranger brushed aside my thanks by saying "The sky is the most beautiful yellow".

Fact. Skylark in 1982
1.At this time our grown son Gary had red golden hair.
2.At that time Gary safely retrieved my bright yellow Skylark from a perilous height bathed in brilliant sunset light.
3.At this time Gary brushed aside my thanks by saying "It's a cracking view from up there Dad".

The seventy feet high tree at the very top of which Skylark became entangled and was rescued by the author's son Gary. A diminutive author at the foot of the tree lends scale. This extraordinary sequel to an event that originated four decades before is portrayed in the text.

Skylark R/C model in first untethered vertical take off mode.

Skylark in static hover mode. The helium lift was just under 10% of the all up weight.

First trimmed forward flight with Gary at the controls.

*Skylark in extensive indoor VTOL and transition to forward flight tests. Note temporary bifurcated inflated wing which was **bright yellow**!*

Skylark has landed. First outdoor VTOL free flight with vertical controlled landing and one very relieved and delighted designer.

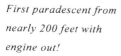

First paradescent from nearly 200 feet with engine out!

Skylark main frame showing cockpit bay, propulsion/lift ducting and undercarriage.

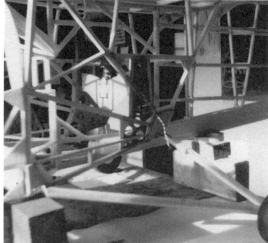

View looking aft showing swivelling engine mounting spat, servo drive mechanism, low mounted fuel tank, and tail-fed spine anti-torque duct.

Close up view looking forward showing mainframe construction, swivelling propeller/engine and forward flight vectoring propulsion duct.

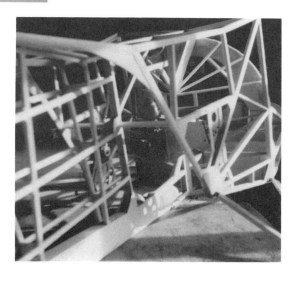

Among all the many lift cells and
paraphernalia , Skylark nearing
completion in the workshop.

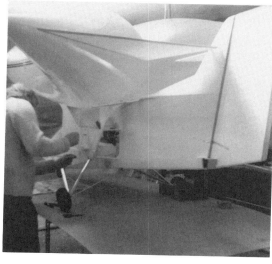

Showing one half of the main
propulsion duct and small tail
anti-torque/yaw control duct
situated at the base of the rudder
hinge line.

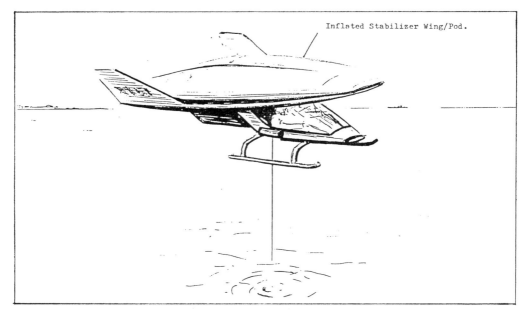

Inflated Stabilizer Wing/Pod.

Full size design of the Skylark hoverplane as a cheap, fast VTOL reconnaissance piloted aircraft or drone for the Navy.

9
The Hoverplane 2ⁿᵈ.Generation

1982`The Guppy'a two plus two seat low skimming,single fan variant.

As with the other variants of the Hoverplane, the full size lift cell would be a delta lifting body configuration with 25% helium and modest wing loading. An engine out from hover situation can be sustained. As stated, the material from which the model stabilizing pods were manufactured was an extremely tough thin plastic sheeting bearing the trade name `Tuftain'. Due to weight problems we had to use this in its single membrane form which is capable of being stretched up to about 40% before rupturing.

In full size application however this basic film is used as a mating solution for the various layers of woven nylon or mylar substrates, which adds to the rip strength, but reduces the stretch. Modern inflatables fabricated from these materials have enormous rip strength and cannot be explosively decompressed as can toy balloons. They can take a bird strike and locally deform before resuming full shape, which is more than the wing leading edges of some aircraft can do. Even a string of bullets does little more than produce a series of half inch long slits in the material, which due to the relatively large contained volume at low pressure and separate internal inflatable air cells, takes considerable time for the gas to escape, allowing a safe controlled descent of a stricken machine.

One that got away

When assembling our models during indoor flight testing it was common practice for us to stow a gas filled pod on the polythene lined ceiling of the workshop by the simple procedure of letting it rise of its own accord. The following day it was retrieved by pulling it down by its dangling tether strings; this proved both expedient and workshop spacewise efficient. Having assembled the model, last minute cg (centre of gravity) trimming was checked and final adjustments were made. This either entailed moving several small lightweight onboard units a comparatively long distance , which could be inconvenient, or moving one heavier unit a short distance, which could be less so. One such item was usually the onboard battery which had to be accessible in order to recharge easily. With the Guppy model this battery finally found its most suitable resting place at the very tip of the stabilizing lift pod, as shown in the accompanying photos.

One very fine day with a cloudless sky found my son and I flying Guppy, when for some reason I decided to dismantle the pod in the field before returning to the workshop. Now, in hindsight, that was an extraordinarily stupid thing to do, because engine oily fingers and a large balloon tugging away with several pounds of lift don't go together very well! In the workshop we become somewhat blase about such slip ups and we had good fun retrieving the thing from the ceiling, but in a very large open space......On that particular day one moment I was restraining the equivalent of one very oversized helium filled balloon, the next instant it was rapidly ascending vertically like a proverbial rocket! Indeed, relatively heavily ballasted forward as it was, it naturally stood on end presenting only its diameter to the upward flight, thus its drag coefficient would have been about .05. Knowing the frontal area, the effective weight and the gas lift, a few sums later revealed that Guppy's pod - with or without straightening fins, but nonetheless now a highly streamlined missile - would have been ascending at 95ft per sec or 65mph!

Etched on my mind is that moment when my son and I stood in silent disbelief at the extraordinary spectacle of a yellow dot silently and rapidly disappearing into that clear blue sky. Our reactions were a mixture of spontaneous amusement initially tempered by the implicit inconvenience, loss of time, cost etc. There was nothing for it but to continue packing and return to the workshop. Then having worked out the average ascent speed and bearing in mind there was nothing to stop the envelope from expanding due to reducing atmospheric pressure with height, together with the material's 40% stretch limit, I had a good idea as to the time and altitude at which the envelope would rupture, that is when it would split wide open and together with a five ounce battery come tumbling back to earth. This turned out to be approximately five and a half minutes at some 30,000 ft. (or five miles)!

Fortunately the light winds were in the right direction to place the point of impact somewhere out to sea, but I had precious little time to check with the local police, coastguards and not least Central Air Traffic Control, who surprisingly enough registered the incident with generous thanks, but apparently little concern. Now I know its a big place up there, but Guppy's pod was venturing into jet aircraft domain, including Concord, and I was well aware of what an errant equivalent of a five ounce lead weight could do in the intake of a turbojet engine! Needless to say we made certain that such a thing could never happen again.

R/C model of Guppy two seat single fan runabout in states of construction. The bottom right hand picture is not of a yogi contemplation, but the author's son Gary topping up Guppy's stabilizing pod with helium, conveys an idea of its errant free ascent mode.

Guppy's fuselage
nearing completion.
The forward comp-
artment is the R/C
bay.

Showing one fore and aft
divided air duct. In full
size vehicle this would be
bifurcated to either side
of the centrally placed
cockpit. The central streamlined object is the fuel tank.

Aft view showing
details of swivel-
ling single fan duct.
Note the coupled
control surfaces.

Showing Guppy's unique single vectoring propulsion/lift fan and duct and the four sturdy inflatable lift cell mountings.

Checking surface effect lift. Note the use of experimental inflatable sponsons.

Getting ready for vertical lift off. Here again a delta wing planform lift cell would be about half the size of that shown.1

No threshing heli-copter type rotor blades to worry about or risk of explosive decom-pression of the lift cell, as Guppy showed indifference to the nearby trees.

'Look no hands' Our overgrown lawn was just right for Guppy's deep surface effect skimming trials,which in both these photos was over one foot.

A measure of the benign nature of the hoverplane principle is demonstrated here with the sponsor's (Godfrey) Hughes 500 in close proximity.

Above. This overhead shot of Guppy hovering shows the fore and aft jet slots and vectoring control surfaces. The package on the nose of the lift cell is the battery as mentioned in the text.

Balloons, balloons galore, test rigs and models adorn our workshop. After the initial shape lofting the author's wife became very dexterous in helping to seam the envelopes on the curved wooden rails of my upturned favourite rocking chair!

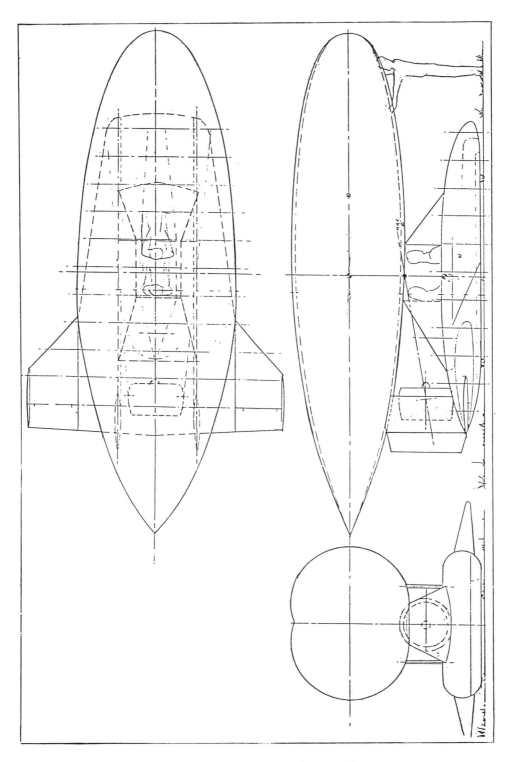

G.A. of the 'Guppy' hoverplane cheap runabout.

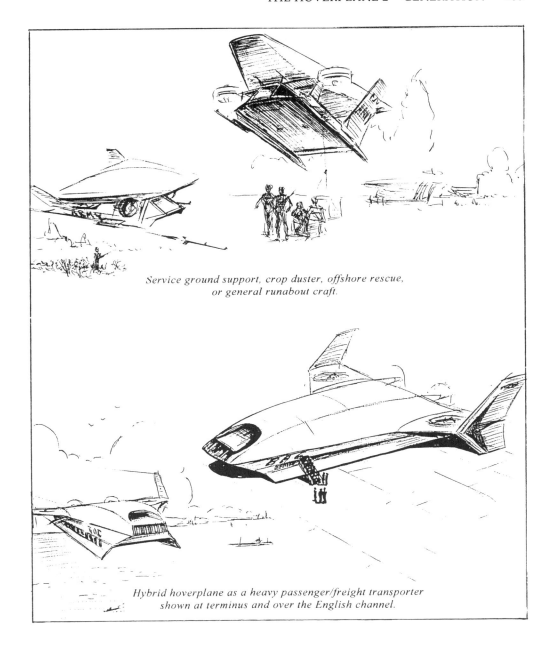

Service ground support, crop duster, offshore rescue,
or general runabout craft.

Hybrid hoverplane as a heavy passenger/freight transporter
shown at terminus and over the English channel.

Even a brief acquaintance with the hoverplane principle reveals that application could be focused on a fairly wide spectrum of size and utility. As the hydrostatic lift compliment can be as low as ten per cent of the gross weight, it follows that the small end of the size spectrum yields fairly modest craft, adaptable for close ground effect, sport and/or tourism. While the comparatively low-tech reflects a correspondingly modest cost bracket. The middle of the size range suggests a fairly wide public transport application, and at the larger end there would be the heavy freight carriers typified by the design studies the author was invited to prepare for B.P. and shown in the following section.

1982. The Workhorse. Rough terrain vehicle study of the hoverplane.

This study for the use of the Hoverplane principle at the 20 and 100 tonne payload size was carried out for B.P.Research by the author.

Operational versatility

A typical Hoverplane mission of some 58 miles is depicted more graphically here which would call on all the operational modes of the machine: 1. Maximum load in close surface effect over fairly flat surfaces with correspondingly reduced power input and consequent increased range. 2. Increased power input allowing deeper surface effect over rough terrain and 3. The means to off load one half of the payload secured in detachable pallets (Thunderbird style) and to operate out of surface effect in VTOL mode over very high obstructions up to several thousand feet. The Hoverplane is extremely manoeuvrable both in surface effect and hovering out of surface effect. Thus it will be able to manoeuvre around a rig in confined spaces in the same manner as helicopters.

Pallet off loading

The ability of the Hoverplane to operate much as a free flying `crane' in VTOL mode is made possible not only by means of the Stability Pillow technique, but additionally by the detachable load capability.

This is achieved by forming the main lower fuselage as an empty box bridging the two main catamaran type sidewalls, in which are housed the lift engines, fuel, crew, etc

Structural components

This Hoverplane comprises four basic units.

`Hard' structure, a load carrying fuselage or main frame constructed to aircraft or hovercraft standards, employing folded/tubular light alloy structural members together with bonded laminates. There is no internal lift air ducting.

Two detachable load carrying pallets of approximately equal volume, similarly constructed as the main frame and contained within it.

Power units and four lift air systems housed on and within the main frame, according to standard aircraft/hovercraft practice.

`Soft' structure, a flexible inflatable wing/pillow manufactured from extremely tough airship type fabrics, formed with integral vertical longitudinal catenary members. Attachment of the pillow to the upper part of the main frame is by well established airship methods, i.e. seamed/welded patches and plates etc.

Two load carrying pallets designed to accommodate one half total load each, form detachable members which employ the fuselage box much as a stowage hangar.

The watertight pallets (which are seaworthy, fitted with land wheels and/or outboard motors) can be winched and stacked one above the other within their `hangar' space.

Equipment and freight is stowed and secured within the pallets to permit easy and quick off load and retrieval.

Thus the detach, transit, retrieve principle related to otherwise inaccessible areas enables the Hoverplane to accomplish what one conventional hovercraft plus one helicopter employing sling conveyance over many trips, plus perhaps one ship, would otherwise be called to do. There can be little doubt as to the cost effectiveness of this new system, moreover it is abundantly clear that operation in some parts of the world make the versatility of the Hoverplane not only desirable, but a dire necessity.

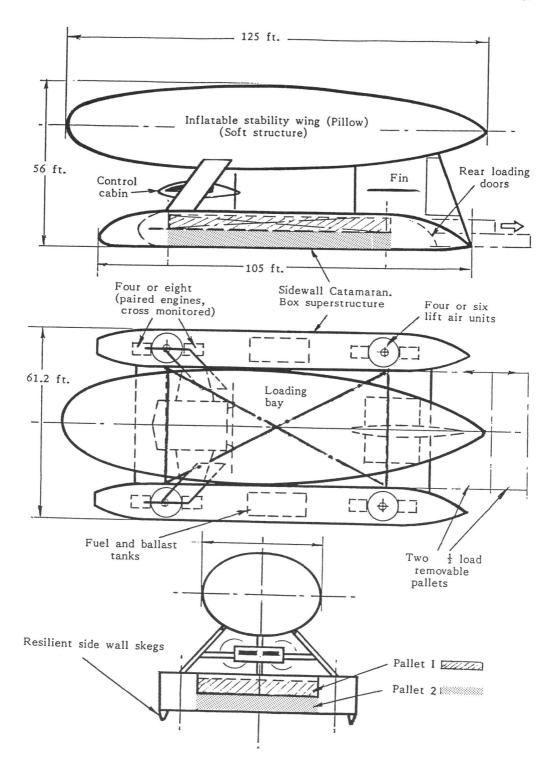

Hoverplane size and structural components.

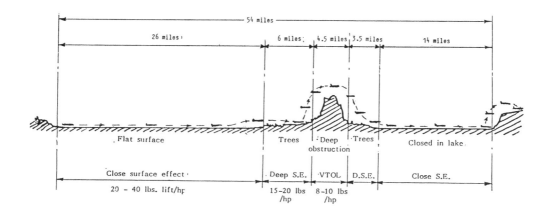

Hoverplane operational versatility.

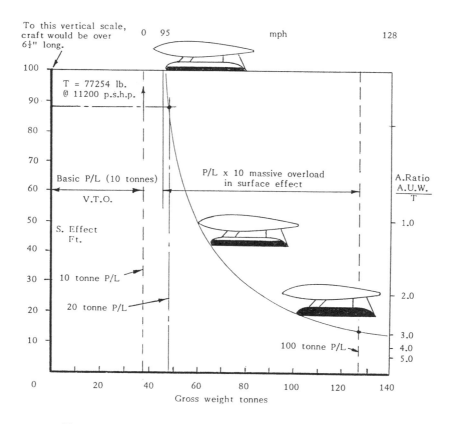

*20 tonne payload hoverplane variation in height for given constant
thrust and increasing multiples of payload from free hover
down to close surface effect.*

Off loading half payload pallet preparatory to vertical lift off and climb over forest range.

Traversing deep fissure in surface effect,

Manoeuvring around rig on confined island touchdown area.

20 tonne payload rough terrain hoverplane operational versatility.

1984. A premature public debut of the hoverplane and an end to the Workhorse?

As previously discussed, in 1982 BP at Sunbury were showing meaningful interest in the Workhorse variant of the hoverplane, as they had the need to transport large heavy loads of equipment to remote parts of the world, in particular Mutluck in Alaska where they were searching for oil. However due to the extremely rugged nature of the terrain which was inaccessible to heavy load carrying vehicles like the hovercraft, and the fact that some of the equipment was too large to be carried by helicopters, the Hoverplane was a possible solution for such a role. BP had already taken the precaution of obtaining verification from Cranfield College of Aeronautics and the Wolfson Unit of Southampton University who both agreed the concept was inherently sound to warrant the building of a full size manned test vehicle.

The remainder of this section is an abridged version of the closing events which were to follow.

BP had offered funding for a test vehicle build which was very acceptable to the author, but there were conditions; they would have to obtain any patent rights before they could proceed. The sponsor objected to this clause and arranged to borrow more funds for us to build the test vehicle ourselves.

Meanwhile BP went on with their explorations at Mutluck and as whimsically portrayed in their television commercial, they didn't find oil, they found water......lots and lots of water! Yet there were other problems in store.

Anatomy of a failure

Having shown the Grasshopper on the Tomorrow's World program, it was vital to get the test rig done quickly. Accordingly Gary and I worked round the clock designing a full size test vehicle, which we could have managed on the available budget had we kep to our original design (shown in this section),but unfortunately the sponsor was persuaded by other people to accept an offer to build the test rig by engineers at a local aerodrome, who were then desperately looking for work. Now it is easy to be wise in hindsight, but they were introduced to me as being the `very best', but later I was to discover they consisted of one very clever aerodynamicist and one equally accomplished technical illustrator. Hardly the best ingredients with which to start on a full size prototype test vehicle!

In view of the short time available however I was prepared to go along with this situation *provided I could remain in control of it.* Yet despite my objections to the sponsor he was overruled by them and they proceeded to cannibalize my original scheme for the test rig and produced their own drawings, thus increasing the time and the cost. As they were more aircraft orientated people totally without experience in surface effect vehicles, they insisted on furnishing *their* version with quite unnecessary flexible skirts, which can in fact be quite a hindrance to stability in deep surface effect due to the constantly shifting centre of pressure. Moreover, despite adequate verbal and written instructions - together with visual aids in the form of drawings, models and photographs - as to the dire necessity of *rigidly* mounting the stabilizing pillow to the hard structure, as will be seen from the enclosed photographs, these bright lads anchored a cheap advertising balloon by *wires* to the structure enabling it to oscillate in all axes, forwards, sideways *and* heave! Also it has to be borne in mind although helium is lighter than air it still contains inertial mass in this particular case amounting to a weight of over one hundred pounds. Thus a sad summary of the position revealed that as *everybody* knows that *Hovercraft* have skirts they had to fit them, moreover as *everybody* knows baskets and gondolas are flexibly suspended from balloons by wires, we got them also!

I objected to all this and they graciously compromised by providing four *one inch* diameter aluminium tubes, and when I pointed out that the ends of these could be lethal they kindly fitted these with table tennis balls `to prevent the tubes from puncturing the balloon', whereas a glance at some of the photographs of our *models* and drawings - with which they were all too familiar - shows quite clearly we used large padded saddle pieces sometimes up to one foot long in which our *model* lift cells were cradled! Moreover I was assured the flexible skirts would be omitted, which they were not.

As previously stated the material from which full size balloons are manufactured is extremely tough often consisting of several sandwiched layers which can weigh anything from several pounds per square yard for large dirigibles down to one and a half ounces per square yard for small advertising balloons. However the sponsor in his desperate need for urgency relied on the questionable expertise of other people who thought they knew best to purchase the balloon and as *everybody* knows a balloon is a balloon, so we got one.....an advertising balloon of one and a half ounces per square yard!

`As high as I can get it'

Coinciding with this worrying situation, despite my son's considerable help over the years, without which we would all have been seriously penalized, he was unceremoniously stood down in favour of the `expert' crew with whom I had been saddled! I thought such indifferent behaviour towards my son together with the sponsor allowing me to be dominated by this motley crew was quite disgusting, said as much and promptly tendered my resignation from the company. A few says later found me attending a meeting in London with a representative of these `contractors', the sponsor, my own accountant and Mike a helicopter pilot friend of the sponsor, who was to pilot the test vehicle at a demonstration, as I thought, to representatives of BP.

Perchance over a rather frosty lunch I sat near to Mike and joined him in conversation about the forthcoming test of the *tethered* machine. I remember the chilling moment quite clearly when responding to Mike's remark about taking the tethers off on this *very first flight,* I said "What do you mean, fly the damned thing first go without them?" I looked incredulously at Mike as I continued "surely not on to the surface effect plateau?" (which could have been ten feet). He responded a little surprised at my serious attitude before he finally said "maybe even higher" I could not believe what I was hearing and by now quite apprehensive I gasped "how high?". With a don't worry smile which I hoped would any moment turn into a mischievous grin he said "G (the sponsor) has asked me to take it *as high as I can get it!"*

Now it is generally known that high altitude balloons that contain helium gas are usually loosely filled at ground level in order to accommodate the interior expansion which occurs as the outside atmospheric pressure decreases with increased height. However, as discussed earlier, in the instance where it is important from aerodynamic consideration to maintain an efficient shape, internal `ballonets' filled with pressurized air - usually supplied by the thrust fans - act as displacement valves which enables the shape of the envelope to maintain fully tensioned integrity. As stated it has always been accepted that variants of the Hoverplane would employ this long tried principle.

However, despite the fact that I had constantly made a point of stressing this, both in written reports and verbally, this crude test rig did not carry this facility, due to the fact that I never intended it to be tested without ground secured tethers anyway, to say nothing about VTOL! It must be borne in mind that the *minimum* thickness for the lift cell *I* had called for was no less that seven ounces per square yard and we had got one and a half ounces! This together with the fact that helium gas expands by one third its volume per one thousand feet gain of altitude......By now I was *very* concerned for the safety of the pilot. For I knew beyond any doubt that with such a set up the cheap advertising envelope was in grave danger of rupturing, and although the engine lift thrust would not be affected, the stabilizing torque would be ruined and in any case the vehicle was not fitted with *automatic* cross monitoring engine control to the four fan units. Thus the chances were someone could have been seriously hurt or even killed. But the most extraordinary thing was, nobody seemed to want to hear me when I had protested at that meeting!

That night I traveled home very troubled, I was greatly concerned for Mike and my own future and everything we had worked so hard for. Therefore I wrote a strong letter of protest to the effect that the treatment of my son was bad enough, but this situation was intolerable and I had no alternative but to confirm my resignation as chief designer of Airbilt and absolve myself from any responsibility of what might happen.

My resignation was accepted, my protest was ignored and the *public debut* went ahead. So, unbelievably on the *very first flight* not only was the machine untethered, but the national daily press *were invited* en masse, a more effective self-destruct of a company is hard to imagine! I had washed my hands of the whole affair and did not attend this fiasco, but I prayed for Mike's safety, and the rest is history. But the true account of that day is this. Due to the totally inadequate floppy attachment of the cheap balloon to the hardware (as is adequately shown by comparing my photos and drawings here with the contraption photographed by the national press that day) as most engineering students could have predicted, the initial bouncing motion of the never before tried vehicle against the inertially unrestrained balloon, in spite of the novel ping pong ball *protective* `end pieces' instantly thrust one of the `support' tubes through the too thin balloon skin. Whereupon in order to prevent serious loss of helium, Mike brought the craft down where a hasty patching job was done with sticky tape, no doubt to the amusement of the onlookers and press! But the fiasco was not yet over because reserve helium was unavailable, so the sponsor sent off for an industrial air blower and topped the depleted balloon up with that!

Now as most students are aware, air and helium do not readily mix, it has to be stirred a little, with the result that not only had the vehicle begun to get too heavy for the little motors, but the whole thing was now badly out of trim. Therefore Mike had little alternative but to put the pair of motors at the lower end `through the gate' (maximum hp rating) with the immediate and dramatic result. A piston rod shot straight through a cylinder wall, shearing off a propeller blade which mercifully missed the onlookers. Nobody was hurt, except my pride, and I wasn't even there to `enjoy' that. Copies of some of the news items are shown here. Incidentally the rights of the Hoverplane were *never bought from me* as stated by our `publicity manager' to the Western Daily Press. However over the years such disappointment and disillusionment has been largely overshadowed by the sense of achievement, hope and excitement, and we enjoyed our association with the Hoverplane.

Even so, this close encounter with the dire need for oil and all that that implies, has now become one small cog in a train of events which has given me an insight into our questionable competence to handle a momentous decision that assuredly one day mankind will have to take. Therefore it cannot be over emphasized, it is in the wake of this host of sometimes dramatic, political, industrial and bizarre social experience lies the origins of my decision to publish my ultimate findings in my book The Cosmic Matrix.

Students of the paranormal may be intrigued to note that this was not the first occasion when a piston went through a cylinder wall. As quoted in Chapter 4 the same thing happened in Wales in the car in which I was travelling in 1962 and both instances involved *crucial public demonstrations* of my work. Moreover, although I live on the Isle of Wight, both of these events occurred on the mainland and *within a short distance of each other.* That....... is synchronicity!

One tenth scale model of the Workhorse grounded and in free flight, as shown on BBC South and ITV in 1983.

Full size manned Workhorse test vehicle built for BP UK evaluation. First flown in October 1982 equipped with advertising quality material lift cell and quite unnecessary flexible skirts. Note there is a complete absence of vertical stiffening between the lift cell and the hard structure, necessary to negate inertial loads.

FLIGHT OF FANCY: Captain Mike Smith ready for take-off in the Hoverplane.

Oh, those magnificent men . . .

WESTERN DAILY PRESS
21ST MARCH 1984

By Peter De Ionno

THE Hoverplane — part hovercraft, part aeroplane, part balloon — took to the air on its first manned flight from Bristol Airport yesterday.

And Mr John Riseley, a director of Airbilt Ltd, which has developed the craft said: "It's a reality now.

"It was a gamble until we flew one with a man on it."

The Hoverplane has taken 12 years and cost £100,000 to develop and will cost even more to put into production.

It was invented on the Isle of Wight by Mr Leonard Cramp, but Airbilt, which has bought the rights, has based its research and development at Bristol Airport.

"It's basically a hovercraft which can clear obstacles, but there's no reason why it cannot fly to 10,000 feet," said Mr Riseley.

"It rides on a column of air from downward-facing engines rather than a hovercraft which is supported on a cushion of air trapped in a skirt.

"When it is airborne, a delta shaped envelope above it will give it lift."

Mr Riseley said the machine used a gas bag to give lift which gave it the appearance of a barrage balloon.

Its cost could vary from £25,000 for a two-seater "sports" version to many millions for a large-scale "flying crane".

"We think it could have great potential as a vertical take-off and landing inter-city shuttle," he said.

Mr Riseley said small versions could be in production in 15 months.

Today the plane will be put through its paces for a major oil company, which is interested in a version which can carry loads of up to 100 tonnes.

"The Government hasn't shown any interest at all but British Aerospace are interested and so are one or two others." said Mr Riseley.

Examples of how otherwise well meaning sponsors, inadequate `publicity managers' and the press can make a good job of totally misrepresenting a product and its inventor.

Hoverplane's spot of bother

BUT THEN, EVEN THE WRIGHT BROTHERS HAD SETBACKS

A NEW flying machine called a Hoverplane made a disappointing debut at Bristol Airport yesterday.

It was a case of more bother than hover for the prototype machine which is part aeroplane, part hovercraft and part airship.

After nearly two hours of trying to get it more than two feet off the ground, Captain Mike Smith, who normally flies helicopters, abandoned the first public demonstration flight.

Mr John Riseley, a director of Airbilt Ltd which is based at Bristol Airport and is seeking financial backers to build production models, said: "Of course it is disappointing but then it is all part of aviation history. Even the Wright Brothers had many problems before success."

He said one of the Hoverplane's four engines had failed. "It could have been embarrassing but we were able to demonstrate our ability to fly on three engines. It's like getting a puncture in a car."

Airbilt sees the Hoverplane's future as a versatile transport aircraft capable of carrying 1,000 tonnes and reaching heights of 10,000 feet.

Doomed balloon ... the prototype before take-off yesterday.

WESTERN DAILY PRESS – 22ND MARCH 1984

Oh bovver, our hover won't fly!

NEW hoverplane had its maiden test flight yesterday and ran into a bit of bovver.

The machine, part-plane, part-hovercraft and part-airship tried to fly for two hours at Bristol Airport before tests were abandoned.

The makers, Airbilt Limited, who spent £100,000 and 12 years' research, say they will keep trying.

THE SUN 22ND MARCH 1984

Hoverplane bother

THIS new flying machine, above, called a Hoverplane, made a disappointing debut at Bristol Airport.

It was a case of more bother than hover for the prototype machine which is part aeroplane, part hovercraft and part airship.

After nearly two hours of trying to get it more than two feet off the ground yesterday Captain Mike Smith, who normally flies helicopters, abandoned the first public demonstration flight.

Mr John Riseley, a director of Airbuilt Limited, based at Bristol Airport, is seeking financial backers to build production models.

He said: "Of course it is disappointing but then it is all part of aviation history. Even the Wright brothers had many problems before success."

The man-carrying prototype was produced after 12 years' research and some £100,000 investment.

Airbuilt sees the Hoverplane's future as a versatile transport aircraft capable of 1,000 tonnes and reaching heights of 10,000 feet.

The prototype aircraft comprises a cigar-shaped silver balloon, under which is suspended a platform carrying the pilot and four downward thrusting engines.

BRISTOL EVENING POST – 22ND MARCH 1984

HOVERPLANE'S DEBUT FLIGHT ABANDONED

A prototype of a new flying machine called a "Hoverplane," which is part aeroplane, part hovercraft and part airship made a disappointing debut at Bristol Airport yesterday.

After nearly two hours trying to get it more than two feet off the ground, Capt. Mike Smith, abandoned the first public demonstration flight. The design is backed by Airbuilt Ltd.

The "hoverplane" comprises a cigar-shaped silver balloon, under which is suspended a platform carrying the pilot and four downward thrusting engines.

DAILY TELEGRAPH 22ND MARCH 1984

Bother with hover

A PROTOTYPE 'hoverplane' got less than a foot off the ground when a propeller sheared off during its first public test run at Bristol Airport.

DAILY MAIL 22ND MARCH 1984

*As shown in the accompanying photos and emphasized in the text, this machine as flown was incorrect in a vital area, and without tethers positively dangerous to fly. Yet despite the author's protests and oft repeated efforts to explain this to the sponsor, the **public** demonstration went ahead with the predictable results. Thankfully nobody was hurt, though we could have done without such adverse publicity. Having regretfully already resigned from Airbilt for this and other reasons, fortunately I had not attended this sad end to an exciting adventure which between us had cost us a good deal of money, time and much work.*

The author welcomes this opportunity to state that some of the statistical quotations in the above reports are misleadingly inaccurate.

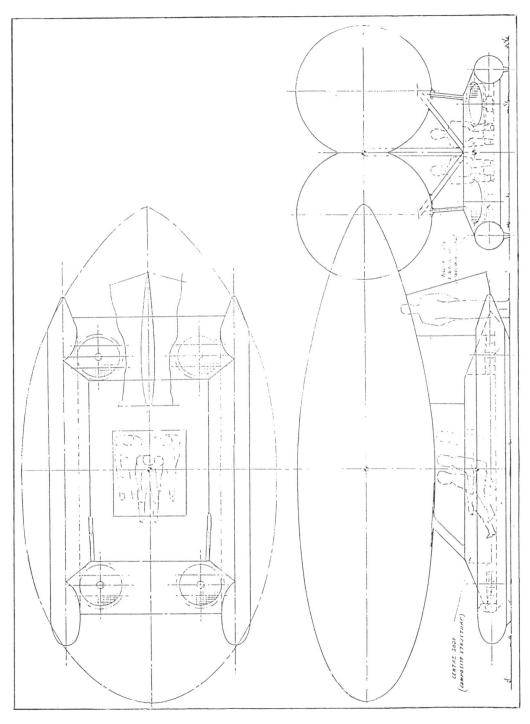

Original G.A. of the test rig as designed by the author. Note the
fixed *side sponsons and* ***substantial*** *pod mountings which are*
vital for the function of the hoverplane principle.

Above. This model was powered by four powerful geared motors and rotors situated at the four inner corners of the hard structure and load bearing boxes.

Showing the twin fifteen feet long lift cells for this R/C model of a Workhorse variant which was the last research programme for the sponsor and BP conducted by the author.

A belated postscript

Although normally one would refrain from capitalizing on other's misfortunes in order to score a point, in this instance I feel sufficiently incensed to state that for me much of the foregoing mishandling by comparative amateurs has been sadly reemphasized by the shocking news concerning the plight of millions of people, men, women, children and their livestock, in the otherwise inaccessible flooded areas of Mozambique. With much due respect to the exemplary crews of a few expensive, limited capacity helicopters, in deep surface effect the low tech relatively cheap Workhorse could carry well over 300 people!

In order to get some facts into perspective, recently I have been reminded of a lunchtime meeting (during which I had tried patiently to assure a lay person about the availability and cost of helium gas) by the rumour that funds are being sought by a large consortium in Germany to build the largest 100% displacement airship ever, apparently with the chief directive of obtaining the world's speed record for airships. The 10% displacement Workhorse would use a mere fraction of the helium gas used by such a playtime colossus!

1971 - 1984 The Channel Wing Cascade Hoverplane. An adaptation and improvement of the channel wing concept by providing a variable cascade for vectored thrust and marrying it with a twin cell layout similar to the Humming Bird.

This machine would also have `engine out' free vertical paradescent capability and fly at over one hundred and sixty knots.

It is perhaps pertinent to note that had the designer of the Edgely Optica aircraft been aware of the channel/cascade vectoring system it would almost certainly offered the Optica VTOL status, or at the very least STOL (short take off and landing). In fact many people mistook the Optica for the author's channel wing version which preceded the advent of the Optica by nearly a decade. Indeed the following model was demonstrated at BP R&D establishment at Sunbury, UK in 1982 before the Optica flew.

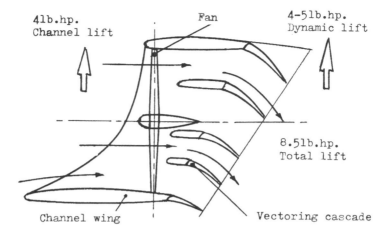

Schematic layout of channel/cascade lift system.

*VTOL channel/cascade hoverplane minus Humming Bird type wing
tip lift cells. Showing channel and deflector cascade.*

Forward view of channel test rig.

R/C model cockpit and duct structure.

R/C model channel wing hoverplane assembled.

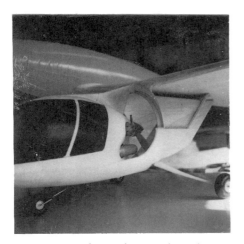

Rearward view showing channel and fan duct.

Forward view on cascade bay.

By marrying the channel cascade dynamic system with hoverplane type lift cells, a total lifting capacity of around 10 - 15lb.hp. can be generated. This offers a very attractive comparatively cheap contender for smaller helicopters which normally yield lifts of around 8lb.hp.

General view showing channel, fan duct and wing tip hydrostatic lift cells.

Rear view of channel lift test rig.

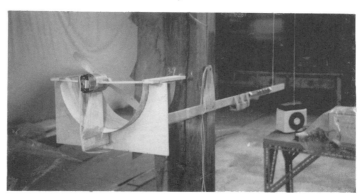

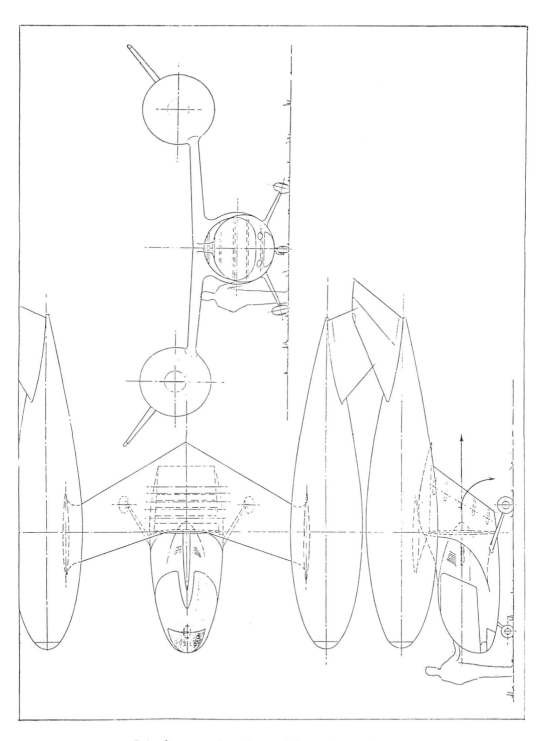

*G.A. of two seat channel/cascade hoverplane with wing
tip lift cells.*

1981 The Grasshopper four seat variant of the hoverplane. Verified at the College of Aeronautics, Cranfield Institute of Technology in 1982.

The one third scale radio controlled test model shown overleaf was aired by the BBC Studio One, Tomorrow's World program on February 24th 1983.

This program reached millions of viewers world wide, both local TV Channels, BBC South and ITV gave it special feature news coverage, video tapes of which still remain. Among the viewers who responded with keen interest were representatives from ICI, the British and overseas Navies and various industrialists.

Sequel to a 26 year old
precognitive dream

For convenience, the Grasshopper Hoverplane was an electrically powered RTP (Rotary Tethered Pylon) model. This system is the same as that originally employed with early models of the Harrier Jump Jet aircraft. Among other things it is easier to `fire up' an electric motor than it might be to start a small IC engine on a cold day, and this sometimes in order to get a few seconds of information. That, and as every modeler knows, a noisy IC engine is not exactly suited to venues such as the BBC's Studio One!

On the Grasshopper we used two amidships swivelling electric motors and fans and in order to counter balance the weight of the umbilical electric supply cables (which tend to drag a hovering model towards the centre of the circuit) it was necessary to offset this by very slightly tilting the whole model outwards. This can be achieved by either asymmetrically trimming the model with ballast weight and/or adjusting the rpm of one motor to give more or less thrust. On one occasion in a moment of `murphy's law) demonstration panic, it had been difficult to attempt either of these refinements, so there was only one other way i.e. bleed off some thrust by finger tweaking the aluminium anti torque blade tabs a little while Grasshopper was airborne. So Gary took the model up to hover as about six feet while I slipped underneath and reached up to do my `tweaking', exactly as I had done in my dream twenty six years before! But how could I have known in those early days that when this event took place there would be such a moderate unpolluted air stream and noise?

The demonstration on the Tomorrow's World program went off satisfactorily with Gary unconcernedly flying Grasshopper in enormous circles over the bemused camera men's heads to say nothing of their terribly expensive equipment! The following day our telephone didn't stop ringing, everybody wanted to own one of these new aircraft which `anyone could fly' (this was a media expression, not ours. Though they weren't so very far out at that).

At the time of writing it is with sombre reflection that I am reminded of a recent TV program in which `cars of the future' was referenced. In this item a very celebrated individual was interviewed during which he presented `a brilliant new idea for future transport'. It was no more or less than a variant of the Hummingbird Hoverplane...... C'est la vie!

Daily Mail, Saturday, February 26, 1983

TV Mail

TELEVISION REVIEW BY MARY KENNY

Variety

It is important to look in on the programmes one's children watch, and I find Tomorrow's World most instructive and put together with a fine sense of pace : there is plenty of variety and yet it doesn't have that hurried feeling that programmes can acquire when there is too much data.

This latest programme had an item about how a money-changing machine works, the muscular composition of a racehorse, accident prevention on drilling-rigs, the hormone patterns of termites, a swimming device to help the disabled, and a modified hovercraft that can really fly.

Much knowledge in 30 minutes, indeed.

Apart from the provisional over sized lift cell, this one third scale radio controlled model of the Hoverplane was demonstrated on the BBC Tomorrow's World program in 1983. The two swivelling propeller ducts are omitted.

Gary propares yet another one of many electric motors for test rig evaluation.

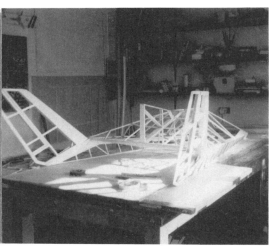

Forward view of fuselage. The pilot's view from Grasshopper's cockpit was envied by many helicopter pilots.

Rear view of fuselage and software support/fins nearing completion.

An electrically powered motor and fan unit under static thrust test.

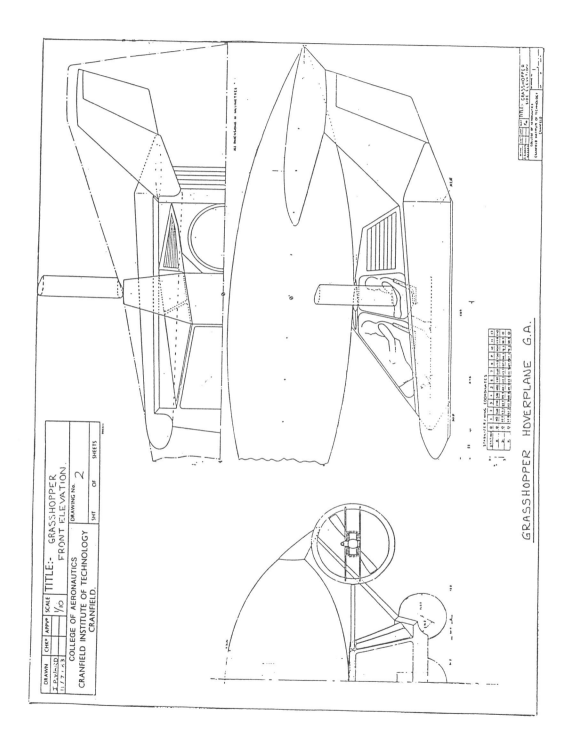

Grasshopper hoverplane G.A. prepared by the College of Aeronautics, Cranfield Institute of Technology, based on the author's original.

This painting by Peter Thorne shows the definitive 2+2 'Grasshopper'
hoverplane verified by the Cranfield Institute of Technology in 1983
as shown on the BBC south and ITV local programmes.

10
The Hoverplane 3rd.Generation

1954 - 1955 The `Aeroskimmer'. The natural development of the `Torus' a true flying saucer shaped vehicle.

This particular variant of the hoverplane now under further development has VTOL capability with an operational ceiling comparable to the helicopter. With a diameter of some fifty feet, it can hover and operate at approximately that same height as a hovercraft in which mode only a hovercraft's pilot's license would be required.

Its useful roles are as varied as they are many, from an on the spot field ambulance with a seven feet high operating theatre, to a luxurious mobile home for four people. It is amphibious therefore it can be based on water or land. There are three alternative methods of providing dynamic lift for the Aeroskimmer, all with a high redundancy rating depending on the specific role envisaged.

The first system employs several coaxially mounted independently powered centrally placed fans, distributing ingested air to a Cockerell type peripheral jet. The second alternative dynamic lift system takes the form of strategically placed light weight engines and fans cross monitored and spaced around the vehicle's perimeter.

The third system offering the maximum power loading, constitutes two contra- rotating Couzinet type peripheral rotors turning at very modest speeds of the order of 30 rpm. This lift system provides a very quiet arrangement of such a high power loading, that together with the maximum helium gas lift compliment, plus surface effect at around fifteen feet, the vehicle could be operated by electric motors and off the shelf batteries for one hour's duration. Moreover the large rotors are so designed as to be able to double up as wind power generators, thereby offering a useful duration extension when operating in this mode.

Of which dreams are made

The Skimmer Mk 1 Hoverplane was initially conceived as a result of the author's all too brief sojourn with a very nice medium class trimaran yacht called Humming Bird (from which the Hoverplane of that same name drew its title). This little craft sported a cabin with galley and twin bunks which occupied an inboard portion of the sponsons support wing. Apart from owning and sailing the boat, I found it extremely pleasant to spend the odd night on Humming Bird while moored on the river under a star studded sky. In fact the idea of travelling about in this fashion appealed heavily to the Romany in me. This, together with the ever increasing news about congested roads at weekends and the longing to be free of all my research work and the idea of taking up my home and travel, was a constant beckoning directive to me. By now I knew I had a very attractive solution, but one day I was destined to do something about it. The Mark 1 Skimmer was the result. The up and over hedges concept had found its mark.

I personally would like to have had the opportunity to `sell up', rent a bit of land in the country and live in a Mark 1 Skimmer and I am sure there are many others who would also. Among those I can count Rear Admiral Middleton, past chairman of British International Helicopters, who paid me a visit to discuss a design study for a large version of this hoverplane to carry one or two hundred personnel. The Admiral was very keen, however the rest of the board failed to share his enthusiasm. Regretfully I have to add that I have learned to accept such reticence (no matter how pretentiously it may be cloaked) is a trait in the make up of a heavy proportion of the British people.

From an old photo taken by Keystone Press Agency Ltd. London. Shows Hummingbird with designer/builder Ken Sides at the helm after winning a `round the island' race. Sometimes when it lay moored on the river Yar, the author had admiringly sailed past the boat feeling the name was familiar. Eventually he was able to purchase it and met with Ken. They were intrigued to discover not only were they near neighbours on the island, but had been so at mainland St. Albans years before, where Ken had built the boat naming it Humming Bird after a popular roadside Inn known to them both.

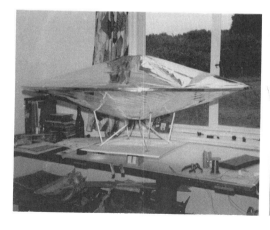

*1954 - 75 Some first steps in the development of the discoidal hoverplane.
Note the use of biconic sections as discussed in the author's article
in Aeronautics 1958 shown in Chapter 8.*

Aeroskimmer Mk.1. hoverplane in maximum surface effect or plateau. Blue underbody is waterborne line.

Electrically powered surface effect mode Lift components: surface effect, dynamic thrust and hydrostatic lift.

Aeroskimmer grounded, but it can equally be waterborne and operate as a boat.

Showing flight deck
and one of four
auxiliary helium lift
cells,

Central lounge, two
lift cells and one
ward room.

Flight deck with galley
diametrically opposite,
central lounge, two
ward rooms and four
auxiliary lift cells.

Upper aerodynamic dome
also functions as the
main hydrostatic lift
cell. The circular
configuration of the
Aeroskimmer offers the
best strength to weight
ratio, greatest volume
for surface area ratio
and requires the
minimum number of
structural parts.

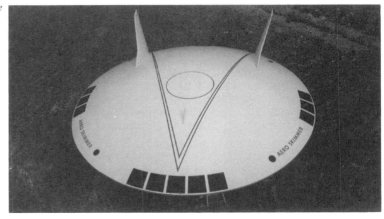

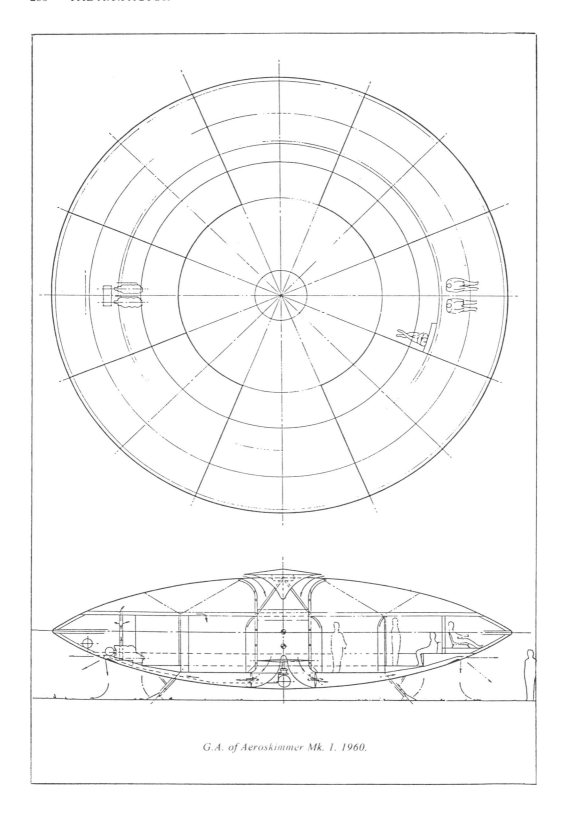

G.A. of Aeroskimmer Mk. 1. 1960.

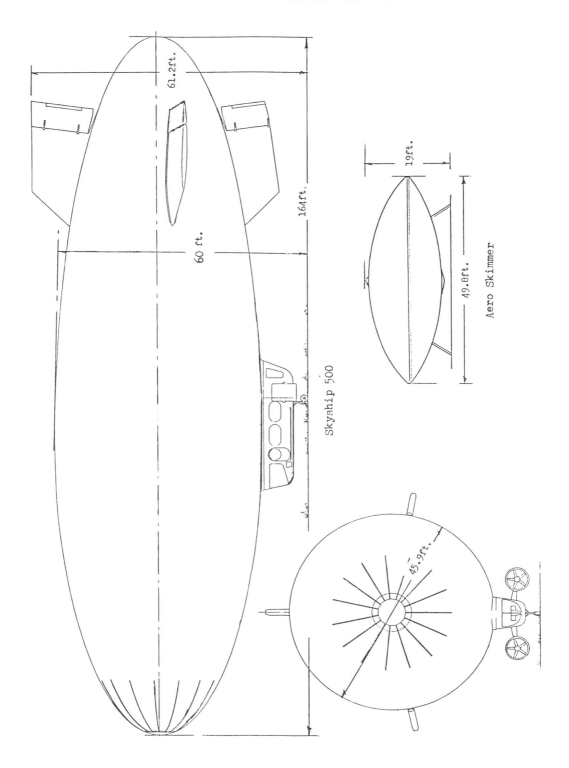

61.2ft.

16/ft.

60 ft.

Skyship 500

19ft.

49.8ft.

Aero Skimmer

45.9ft.

Aero Skimmer Mk.1 to same scale as Skyship 500.

SOUTHERN EVENING ISLAND ECHO

THURSDAY, AUGUST 6, 1987 ——— Southampton — 99th year — No. 30426 ——— PRICE 17p

£½m TO BUILD FLYING SAUCER

1987 A local news item concerning the
Aeroskimmer Mk.1 reflects the sadly laid
back attitude to invention in the UK.

GLIMPSE INTO FUTURE: What the new hoverplane could look like.

A FLYING saucer lookalike could be humming across the Solent inside a year if an Island designer can raise the capital to develop his Disc Hoverplane.

Mr. Leonard Cramp, of Yarmouth, said he has a team standing by ready to build the revolutionary multi-role craft.

The Disc version of the Hoverplane resembles a flying saucer because the domed top half is a streamlined successor to the helium balloon used to lift and stabilise the original Hoverplane concept he designed.

That did not get beyond a crude prototype but Mr. Cramp has now refined the concept to a sleek circular flying wing capable of operating as a surface effect craft or a vertical take off and landing plane.

As well as being hyper-efficient aerodynamically he said the shape had the added advantage of being one of simplest structures to construct to combine lightness and strength.

By MAURICE LEPPARD

Unlike conventional hovercraft it would have a rigid bottom half but a software top to hold the helium.

Mr. Cramp calculates that a 50ft diameter Disc Hoverplane could be developed for less than £500,000 and be ready in six months to a year. The cost of production versions could be brought down to around £70,000.

The version he is planning would be able to carry four people in a VTOL plane role and up to 20 in a hovercraft mode with a speed of about 100mph.

It is not only as a means of versatile transportation that he sees the Disc Hoverplane as a breakthrough. "You can't live in an aeroplane but my idea is that the Disc Hoverplane could be

fitted out as spacious luxury standard accommodation.

"It could appeal to people like Peter de Savary who could use this type of trend-setting craft as a highly mobile and flexible base."

As well as a hoverhome Mr. Cramp envisaged the flying saucer lookalike being fitted out as a flying ambulance or field hospital.

*1987 A local news item concerning the Aeroskimmer Mk.1 reflects
the sadly laid back attitude to invention in the UK.*

1965 - present. Aeroskimmer Mk.2 variant of the hoverplane with resident road/surface effect tender.

This multi engined version of the Hoverplane houses its own VTOL ground effect runabout, power units of which, if necessary, can contribute to the mothercraft lift system. When water borne, under water viewing is available via the large centrally placed viewing aperture in the centre of the floor.

Plans of this machine sent through the mail mysteriously never arrived at the office of a well known UK wealthy entrepreneur who had expressed interest in building a prototype. In fact the circumstances associated with it were almost a carbon copy of the Mattel Toy Rocket incident presented in Chapter 12.

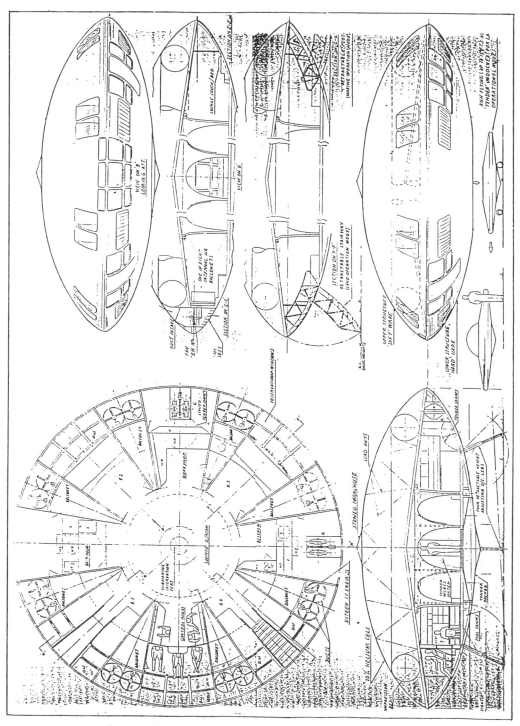

G.A.Aeroskimmer Mk.2 with detachable surface effect tender.
Early copy of the plan which was 'lost' in the post.

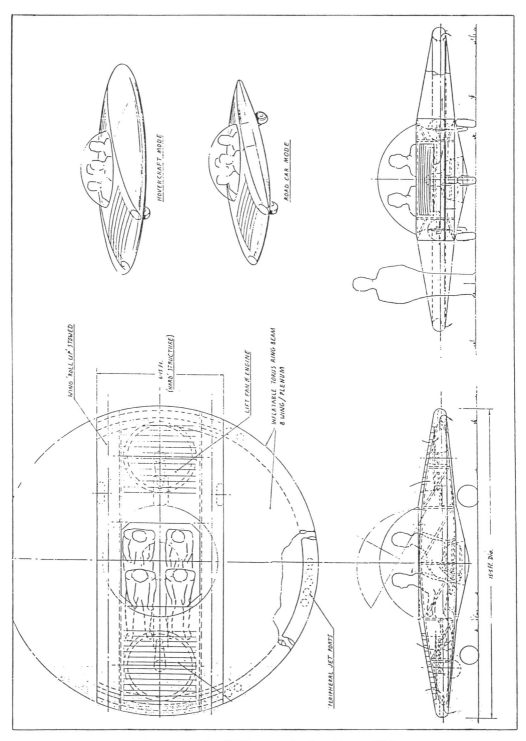

G.A. surface effect VTOL tender for Aeroskimmer Mk.2.
Early copy of the plan missing in the post.

1965 - present. Aeroskimmer Mk.3. variant of the hoverplane fitted with high solidity circumferential rotor/fan.

In this version of the Hoverplane a circumferentially placed Couzinet type rotor provides dynamic and surface effect lift, supplemented by helium lift of approximately 30% of the a.u.weight. Therefore the total effective power loading can be in the order of 100 lb h.p.

This machine could remain in surface effect lift at approximately one vehicle diameter height for one hour electrically powered and with no more noise than a family car.

It is interesting to bear in mind the lost plans of the Aeroskimmer Mk 2 in terms of the storm damaged look alike Spacial MLA - 24 - A in 1985 shown at the end of this section. Note this craft of similar shape was intended to be nothing more than a 100% displacement *airship* , as were the French Atlas airship of 1969 and the British Skyship of 1973. As was shown in Chapter 7, the merits of this discoidal shape were amply publicized by the author as long ago as 1958 in the article in `Aeronautics' titled `The Disc Wing Aircraft'.

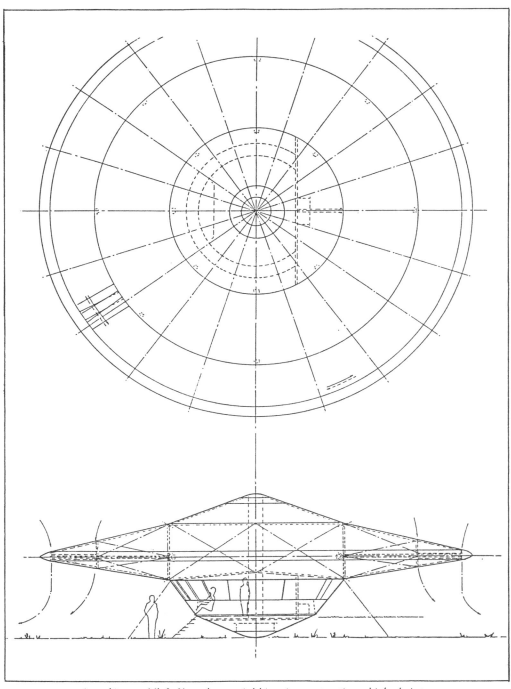

Aeroskimmer Mk.3. Note the special biconic construction which obviates double curves yet produces an aerodynamic shape as shown in the author's article in Aeronautics 1958. This machine will hover in surface effect for over an hour entirely electrically powered.

1987.The Rotorcraft VTOL/surface effect vehicle with circumferential high solidity rotor/fan.

This machine is very similar to the Aeroskimmer Mk 3 Hoverplane with the exception it does not contain helium gas as a lifting stabilizing component. This design was initiated in order to accommodate areas of the world which although having the use for such vehicles, helium gas might be unobtainable. The concept employs the identical aerodynamic lift system as the Aeroskimmer Mk 3 Hoverplane in that it has slow rotating large surface area, circumferential, contra rotating rotors, augmented by surface effect, which - despite the loss of gas lift - can yield high power/weight loadings of around 90 lb h.p. This together with the inherently light weight structure of the disc, can still produce extremely useful hover heights in surface effect around one diameter. Due to the loss of helium lift however, more powerful engines are required, in addition to which, the inherent stabilizing torque is lost, requiring a more sophisticated stabilizing system.

The Rotorcraft would have accommodation for four people and would have a higher top speed than its Hoverplane counterpart. The two and four seat variant of the Rotorcraft are no less interesting in that they have useful surface effect hover height typically of the order of ten to fifteen feet. Also although diminutive by comparable conventional standards, these smaller machines have a large wing area in terms of their span. A factor which is stressed in the Aeronautics article.

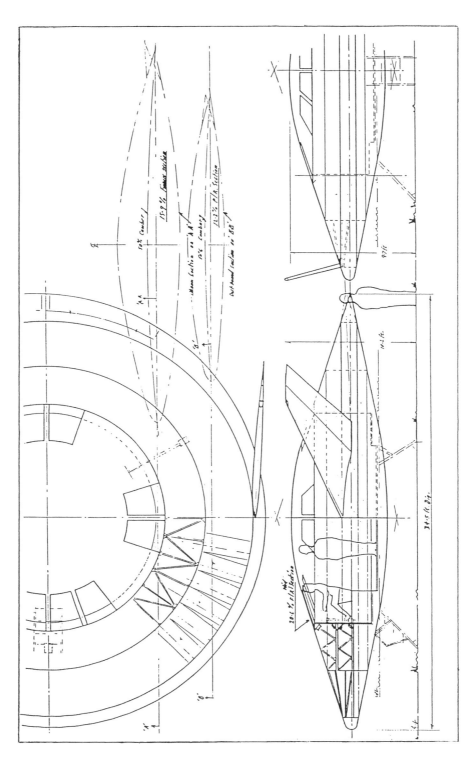

Mobile home VTOL discoidal rotocraft.

Spacial MLA-24-A prototype, nearing completion except for covering on lower half of the elliptical hull

Destroyed in a tornado, the Spacial MLA.24.A 100 per cent displacement airship 1973.

Tested at Cardington, John West's Skyship model of a 100 per cent displacement airship in 1975.

Testing in the wind tunnel the french Atlas 100 per cent displacement airship 1970.

Precognition apart, it seems as though futuristic ideas must have time in which to mature.

An application of the hoverplane as a means of future road traffic easement.

It is common knowledge that even the briefest acquaintance with motor vehicle flow statistics on roads around the world reveals that they are rapidly becoming over populated, particularly on most motorways. It is also known that the motor vehicle is not only one of the worst pollution contributors to global warming, but it is becoming the most serious threat to our limited energy reserves. It is known that a high percentage of fatal accidents on motorways is caused by drivers falling asleep on long boring journeys. And it is known that not only fatalities, but serious injuries to people adds significantly to national budgets running into many millions of pounds a year. The medical treatment and continued support they need is often very long and sometimes requires even permanent after care.

A possible environmental and
cost effective solution

1. There can be little doubt that as things are at present, motorways have to increase in size and numbers , with the attendant building, maintenance and environmental cost incurred.

2. Despite good intentions, railways may be improved, but in the UK they are unlikely to be meaningfully extended. In their present form they are useful, but there can be little doubt there is a limit to building and refurbishing more and more permanent ways and power lines.

3. Faced with the fact that surface transport systems cannot be greatly improved, it might be profitable to even consider air travel. But obviously as the state of the art is at present, this mode of mass transportation is equally untenable. For instance we have seen how there is a limit to the size of helicopters to say nothing of their cost). Both large and small conventional aircraft require supportive infrastructures such as runways and airfields, in open parts of the country from which to operate.

4. Because they can take off and land vertically, helicopters do not require airfields, but their minimum operating altitude across open country is restricted to some five hundred feet. Airships as heavy bulk carriers have been considered, but as we saw earlier the amount of disposable load to size ratio - to say nothing of the supporting operational and maintenance facilities - renders such colossal craft impracticable as far as mass transportation is concerned.

The case for a public
transport system

These are some of the more obvious criteria and of course there are far more too lengthy to contemplate here. Suffice it to say that even lay people are voicing the view that *some kind* of public transport system will eventually have to be considered for the not too distant future. It is known that far more fuel per passenger-mile is expended by using motor cars and small vehicles generally, than it is to move many passengers and/or freight in larger vehicles as is demonstrated by railway systems. For example, by transporting say two hundred people on a given journey at a given speed would represent a significant fuel saving of something like 80%, which in terms of the many millions of road miles covered by the travelling populace world wide each year, would represent a major contribution to reduction in atmospheric pollution, cost and saving in our energy reserves.

So the question has to be asked is there some kind of environmental, cost effective public transport system which in the not too distant future could be available? Not least one which would offer a minimum degree of upheaval.

A hybrid hoverplane as a
public transporter?

Having carefully weighed up the facts for some considerable time, the author has come to the conclusion that the following proposal may have an interesting degree of merit. It is largely dictated by the fact that if drastic change has to come, then we must consider an alternative which would entail the least upheaval. It is suggested that there might be a case for adopting something like VTOL hybrid Hoverplanes as spaced out carriers of restricted numbers of passengers and/or freight, in which role such vehicles would operate at carefully governed heights at about say two hundred feet and immediately over and parallel with, some motorways and existing rail permanent ways. Thus the delineation of such systems would be extended *upwards* to form controlled low altitude air lanes

Proposal for the extension of main railway and motorway routes **upwards** *to provide cheap environmentally acceptable avenues for mass passenger and freight transportation by a sophisticated version of the Hoverplane.*

specifically employed by this traffic.

There would be of course associated problems to consider. For example, contraflowing lane altitudes would be necessary and specific separation from other aircraft traffic etc. Threat to the environment due to noise would be a natural objection, however noise levels of hoverplane type vehicles would be quite benign compared to helicopters, while already accepted motorway and railway noise levels might even be reduced. All transport systems can be adversely affected by weather conditions, including trains running late due to leaves on the lines, while foggy conditions can play havoc on the motorways. Interestingly this proposed type of transport could be equipped with similar adverse weather instrumentation to large airliners. Therefore in this sense hoverplane passengers would have a greater degree of safety than would the fog bound car drivers on the motorways below, while it should also be remembered, hoverplanes can do just that.....hover!

Minimum new infrastructure

The relevant existing railway stations and motorway junctions could, at quite moderate cost, be modelled to accommodate landing pads for arrival and departure traffic, refuelling etc. In fact it might be possible to use the existing systems and reduce their current flow rate by reshaping their functioning, at minimum cost, rather than the horrific alternatives of more and more motorway extensions together with the attendant massive increase in pollution etc.

An appropriate footnote of
the synchronistic kind

At the time of preparing the manuscript for this book in the summer of 1999 an appropriate little incident occurred. I had been in the process of mowing the lawn within a few yards of our workshop (portrayed in several of the foregoing photographs). Nearby runs a country footpath and it is not unknown for the odd paper bag to find its way over the hedge. So when a small piece of paper was disgorged from beneath the mower I barely registered it. But somehow I felt impelled to stop the machine and retrieve the several pieces. In the very act of screwing them into a ball before disposal, my attention was drawn to an inch or so of attached string. I was intrigued and took the pieces to the house and watched Irene piece them together to discover they formed a remnant of a small ticket which had presumably been attached to a kiddies balloon. A written message was barely decipherable but we could read an address to which the remnant should be returned with notification of the finders address. Realizing this was perhaps part of a school fete balloon race we posted them off as requested and promptly forgot the matter.

Returning from shopping about a month or so later, standing on the post box we found a gaily coloured cardboard box with pretty red plastic ribbon and bow simply labeled `congratulations'. Wondering what it was all about Irene took the parcel into the house puzzled by the fact that whatever it contained was extremely light, in fact the box seemed empty. Intrigued she untied the ribbon and out popped a bright multi coloured plastic helium filled party balloon, negatively ballasted by a plastic star tied on to a length of plastic string. Presumably congratulating us for returning the tattered remnant of a winning balloon. Somewhere, some youngster had been delighted that their balloon had won the race.

Now that little balloon had travelled nearly two hundred miles across the south of England, over several miles of sea to the Isle of Wight, and of all the thousands of properties here, it had landed exhausted on our lawn. That same stretch of grass in fact which had been the birth place of many, many expressions of my helium stabilized aircraft....the Hoverplane!

11
Supercars

1940 - present. The CM100 Supercar.

From a preliminary study of a new versatile VTOL aircraft. This work is the natural development of that carried out by the author on the small hovercraft in 1940. Neither it nor the following versions should be confused with the `flying car' being investigated in America, for unlike the CM series, that is not designed specifically to operate as a surface effect vehicle.

Intentionally, and it is felt appropriately, the author has called it the `Supercar', if for no other reason than to acknowledge the predictive nature of its counterpart in the TV series `Thunderbirds'. For it too will go anywhere, it will do anything, it will take off and land vertically, it will be faster, cheaper and quieter than helicopters, which ultimately it could replace. As it may the car, the hovercraft and the boat as we know them, for inherently it is all these things. Simply stated, this new concept can truly be designated an ultimate transport machine of major breakthrough proportions.

The Supercar can take off from water or a medium sized lawn and be parked in an ordinary garage. Moreover it cannot be confused with past relatively crude cars with wings attached, for unlike them it requires no conventional runway or undercarriage.

Although strictly intended for civil application - for which the author feels there could be an extraordinary large market - it is anticipated that the comparatively low manufacture and maintenance costs coupled with compactness, high performance, ground effect and VTOL capability, makes the CM 100 a natural candidate for some specialized service roles.

The CM100 Supercar

Car plus aircraft.

Car plus boat or hovercraft.

CM100 to same scale.

In 1986 Richard Branson sent his technical director Dick Ploughs to view the design study of this vehicle, who was pleased to recommend investigative funding for an experimental surface effect prototype. A budget figure had been discussed and it was proposed that Richard would investigate further on his return with the Blue Ribband trophy. As is now well-known, unfortunately for him his Atlantic Challenger boat struck debris and sank. No less unfortunate for the author because the potentially available funding for the CM 100 was absorbed in the construction of the more successful Atlantic Challenger 2.

As with the previous concepts shown in this book, there has been such enthusiastic response and requests to see more about the CM 100 series, together with the release of the information in my book The Cosmic Matrix, that the author has been tempted to be rather profuse. However this can be shown to be justified due to the fact that although the propulsion units in the series are of the current state of the art (small axial/centrifugal augmented fans) the airframe/aerodynamic format are suited to the lift/propulsion of the future. In other words this is what VTOL aircraft of the future are going to be like. The wing retracting system of the CM 100 is not to be confused with variable geometry aircraft, CM 100 wings have two positions, folded for stowing and fixed extension in flight. Also the tail flying surfaces are semi-retractable thereby reducing the overall length, span and height for garaging. The design provides for a `lifting body' type fuselage together with highly efficient high aspect ratio wings, thereby attaining the best *overall* aerodynamic weight and size compromise.

Stability and control

One of the chief difficulties which has beset VTOL aircraft designs in the past has been overcome in aircraft such as the Harrier, but the degree of sophistication employed places such stabilizing systems way out of cost range for civil application. However the system to be employed in the CM 100 obviates this problem in an efficient comparatively simple manner.

Power units and operating modes

The four *uncoupled* conventional I.C. engines with augmented thrust provide sufficient power for vertical climb at lift off. Static hover with one redundant engine. Controlled vertical descent to surface effect cushioned landing with two or three redundant engines, and height permitting, pull away through transition to normal flight on remaining engine to a controlled conventional landing. Or if surface conditions will not allow, a flair out to vertical cushioned contact. With all engines out and flaps down in a glide mode, the CM 100 will have a stalling speed of around forty five miles per hour.

Maximum range and efficiency

In the max range, max efficiency mode, any two engines can be shut down while the remaining two provide adequate thrust at 65% max cont. rating.

Construction

Although novel in many respects, there is nothing out of the ordinary about the aircraft to warrant anything but accepted methods of construction. However all VTOL aircraft are naturally weight sensitive, therefore it is anticipated the use of modern light-weight, high strength plastics will be employed together with light alloy structure.

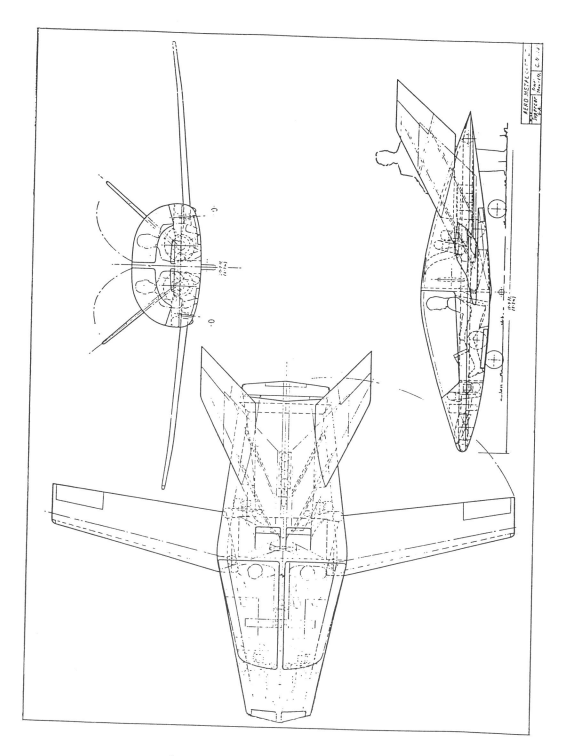

Supercar G.A. with flying surfaces extended.

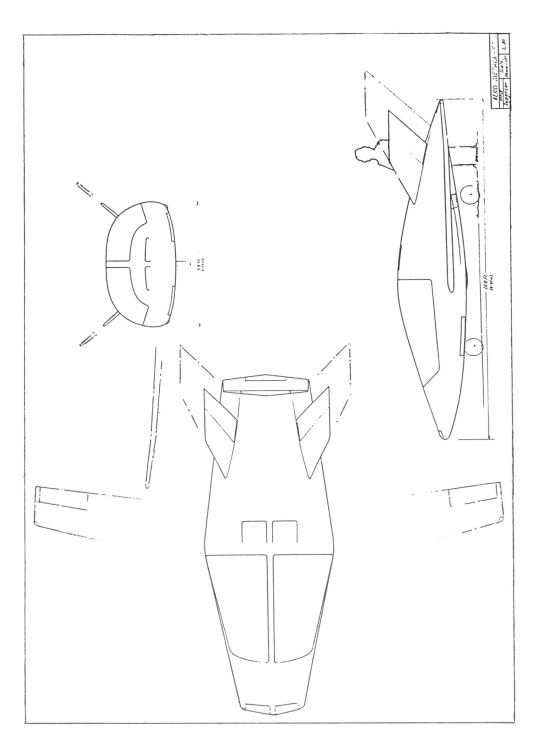

Supercar G.A. with flying surfaces retracted.

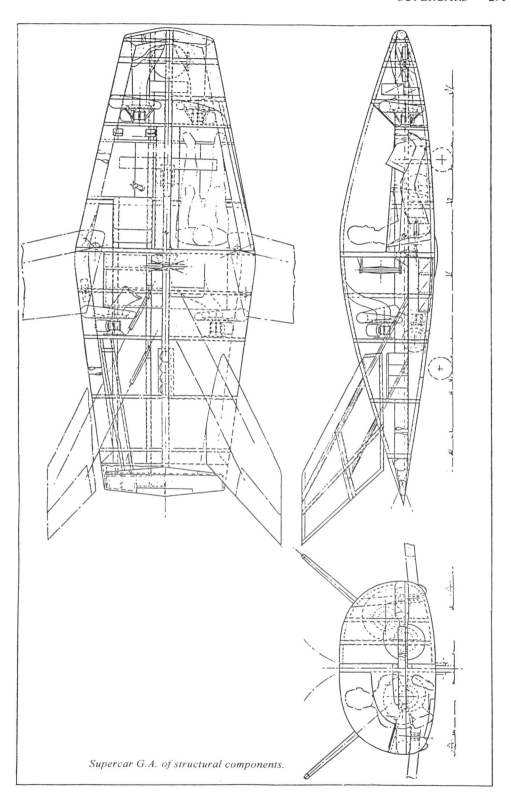

Supercar G.A. of structural components.

CM100 Supercar in road, boat, hovercraft, hover and VTOL modes configuration. Fuselage is designed as a lifting body in hover and/or forward flight.

Wings and V tail retracted. Note, it is a feature of this design that there is an absence of exposed dangerous rotating machinery.

Conventional take off mode or in surface effect with undercarriage retracted.

Note, extended V tail-plane, lifting body fuselage and high aspect ratio wings.

In forward flight, as in surface effect, the generous surface area fuselage contributes significantly to overall lift.

Supercar in conventional flight mode.

SOUTHERN EVENING ISLAND ECHO

MONDAY, APRIL 6, 1987 — Southampton — 99th year — No. 30324 — PRICE 17p

EXPERIMENTAL CAR IN CASH CRISIS

LEN CRAMP shows off his prototype multi-vehicle.

HAVING given birth to the Super 2 light plane the Island has the chance of the baptism of the Supercar. — a revolutionary car-hovercraft and vertical take-off and landing plane.

**EXCLUSIVE REPORT
By MAURICE LEPPARD**

There is a huge cash questionmark though over whether designer Leonard Cramp, of Yarmouth, can convert his exhaustive research and designs into a prototype.

He is reluctantly concluding that unless his search for funds in the U.K. quickly turns up trumps he will be forced to turn to the United States where he has already resisted one "very attractive offer".

If Mr. Cramp's concept of the three-in-one machine conjures images of Chitty Chitty Bang Bang nothing could be further from the truth. His Supercar owes much more to Star Wars than Professor Caractacus Potts of Chitty Chitty fame.

Code named the CM 100, it has a streamlined futuristic look with not a propeller in sight — an eye-opener in more ways than one when production versions have been costed out on a par with a Rolls Royce.

This is because it is powered by four small conventional engines combined with special centrifugal fans — 40 h.p. power-plants offering a

hovering, vertical take-off and forward mode and flight speed of 300mph-plus.

As Mr. Cramp explained it is the design of the fuselage which plays a vital part in making the Supercar such a potentially important development.

It is shaped so that it contributes significantly to the overall aerodynamic lift. Linked to high performance wings and butterfly tailplane the result is a remarkable reduction in weight and size of the hybrid machine.

The CM 100 has a 20.7ft wing span, is 16ft long, a beam of 5.8ft and stands chest high. It can easily be parked in a garage without having to heighten or widen doors. The wings and tail retract into a streamlined shape that personifies a Supercar.

Mr. Cramp (66) is chief designer of Airbilt and chairman of Aerometacraft. He looks on his latest creation as the culmination of a lifetime probing the frontiers of technology, ranging from designing a jet engine in 1953 to his more recent invention of a hoverplane.

He still has an entrepreneur ready to take the hoverplane beyond the test rig stage and at one time had high hopes Richard Branson would be the man he was looking for to pump in the estimated £750,000 to build a prototype of the Supercar.

"Unfortunately around the time I was dealing with Richard the Virgin Atlantic Challenger sank and he decided to concentrate on building another boat. At one period there seemed a good prospect of linking up with a local firm to do it but that came to nothing.

"The irony is there is all the expertise on the Island to build the prototype but time left to find the money is running out fast."

PRESS RELEASE: AN OVERSEAS INTEREST,

VERY LITTLE IN THE UK

US are interested in 'car-hoverplane'

A model of Mr. Cramp's CM 100 "car-hoverplane."

UNABLE to raise money in the Island to launch his latest brainchild, Yarmouth inventor Len Cramp, 66, may take up an American offer to develop a revolutionary car-hovercraft which can also serve as a vertical take-off and landing plane.

Before he can get support from the British Technology Group for his futuristic CM 100 four-seater, he has to convert the present small scale model to a 16ft. span model version of the supercar.

That could take up to £25,000 for the first phase, and although I approached the IW Development Board, who tried to be helpful, it has so far proved impossible to find anyone here willing to put hard cash into a realistic deal," he told the County Press.

At one time he had hopes that millionaire Richard Branson would provide £750,000 to build a prototype — but that was before Virgin Atlantic Challenger sank and Branson's interest turned to producing another boat . . .

Now, with no-one to back him and constant worries after mortgaging his home at Cranmore to keep his ambition alive, Mr.

Cramp has been invited to oversee the CM 100 project in the States for military purposes.

"This is no mad inventor's scheme — it really does work — and it looks as if I may have to take up the American offer," he said.

"It is such a pity — another British idea being exploited abroad . . ."

A skilled engineer, who has worked with Napier's, the aircraft engine manufacturers, de Havilland's and Saunders-Roe, and was a member of the original team behind the Island-developed Turbo Firecracker training aircraft, he left the British Hovercraft Corporation at East Cowes 20 years ago to devote himself to research.

DESIGNS

In the years since, he has produced all sorts of designs, some of which have been patented by major firms, but generally it has been a tale of frustration, disappointment and major expense on a shoestring budget.

"Napier's patented a gas turbine engine of mine back in the 50s, but nothing came of that, and today Rolls-Royce have

come up with the Swallow — virtually a carbon copy — for a projected space vehicle," said Mr. Cramp.

Lack of backing in this country after years of effort also thwarted his other major invention, a Hoverplane combining the characteristics of a hovercraft and helicopter, which left the Cramp drawing board in his garden workshop in 1983.

His elder son, Gary, also an engineer, gave up his job to help Mr. Cramp in the early design work, and the Tomorrow's World programme on BBC television featured the Hoverplane.

"The Americans are now showing interest in that too," he said. "It seems we just don't have the people in this country willing to back new ideas — Chris Cockerell found that out when he designed the first hovercraft!"

The CM 100 is powered by four small conventional engines with centrifugal fans — 40hp units providing hover, vertical take-off and a flight speed of over 300mph.

It has a 20.7ft. retractable wingspan, is 16ft. long, has a beam of 5.8ft., and can literally be parked in an ordinary family garage.

ISLE OF WIGHT COUNTY ORESS APRIL 10th 1987.

Some typical UK newspaper reports reflects the culmination of much time, money and effort.

1984.The two seat CM200 VTOL, STOVL high performance aircraft with augmented surface effect and canard format.

With foldable outer winglets, tandem seating, and maximum width of eight feet, this machine can be garaged and flown off the road. It will be apparent that the format includes the channel wing/cascade as employed in that named variant of the hoverplane. In some roles its performance will match that of the CM 100. But it can also be operated as a fully submersible submarine craft. Some naval personnel have shown interest.

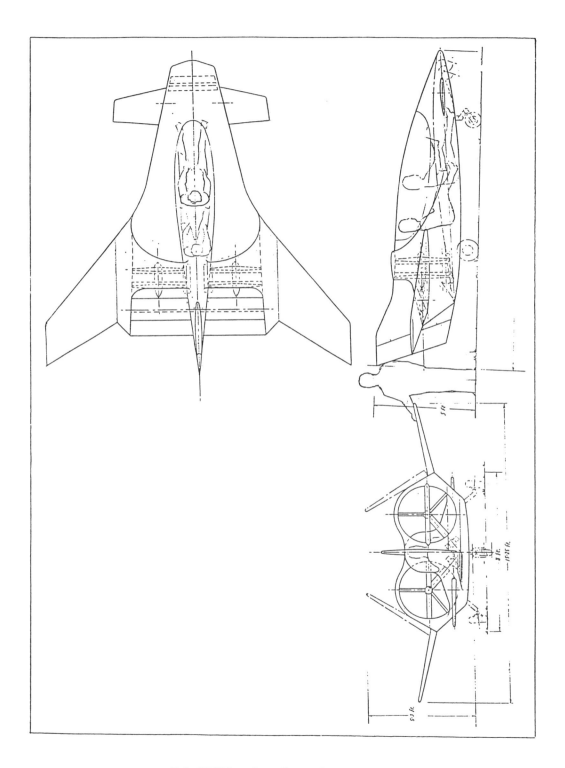

G.A. CM200 surface effect and canard format.

1985. The CMAL 1 & CMAL 2 series. Results of efforts to produce really low cost kit VTOL surface effect vehicles which in essence would be `along the road, over the hedges' type marine craft.

This design allows for both ends of the cost, performance spectrum. But all the variants will be capable of VTOL or surface effect hover at the plateau height of around sixteen feet. Moreover these are vehicles employing off the shelf technology equipment with up to date, low cost stabilizing systems so that most car drivers could `fly' them hands off.

The CMAL 2 can be operated as a VTOL hoverplane by the simple adaptation of the inflatable deltoid wing either with or without helium gas lift. Grounded and fitted with surrounding curtain, this craft would make an excellent camping vehicle.

Due to the inherent simplicity of design which takes advantage of modern plastics and materials, it is anticipated that development time from prototype will be cut to a minimum. Note, the CMAL 1 can also be made to be completely submersible.

CMAL 1 in VTOL, road, hover or surface effect mode. Tail plane and lift/propulsion cascades closed.

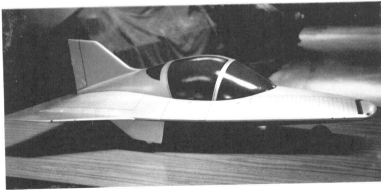

CMAL 1 in forward flight/surface effect mode, U/C extended.

CMAL 1 in submersible mode.

CMAL 1 in waterborne mode.

*CMAL 2 in off road configuration with inflatable wing attached and furnished with
a plastic curtain, will offer a very useful overnight stay or camping facility.*

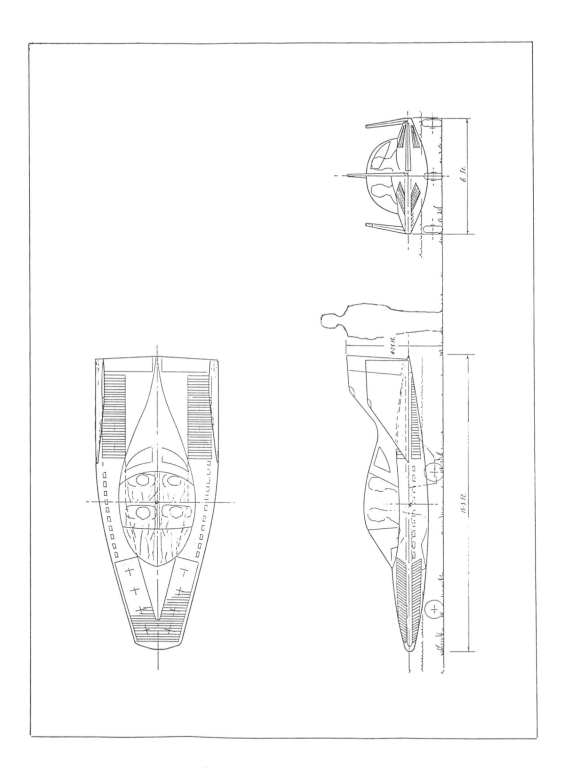

G.A. CMAL 1 tail plane retracted.

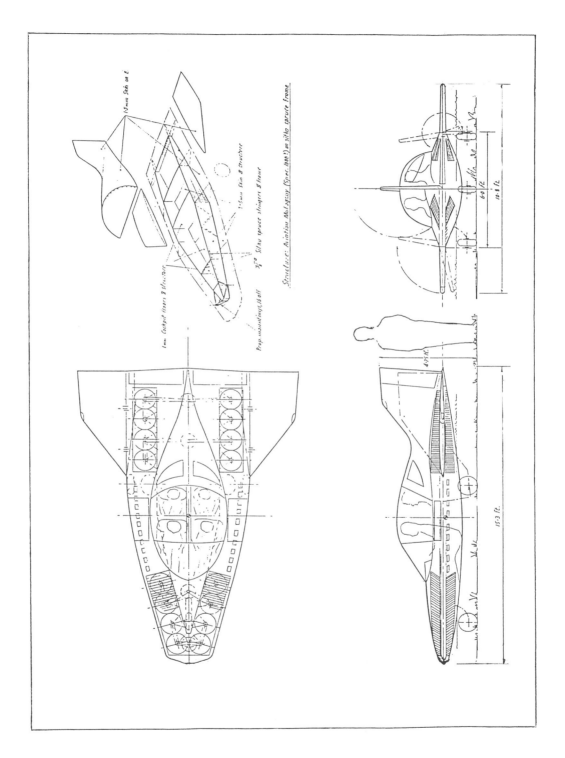

G.A. CMAL 1 tail plane extended.

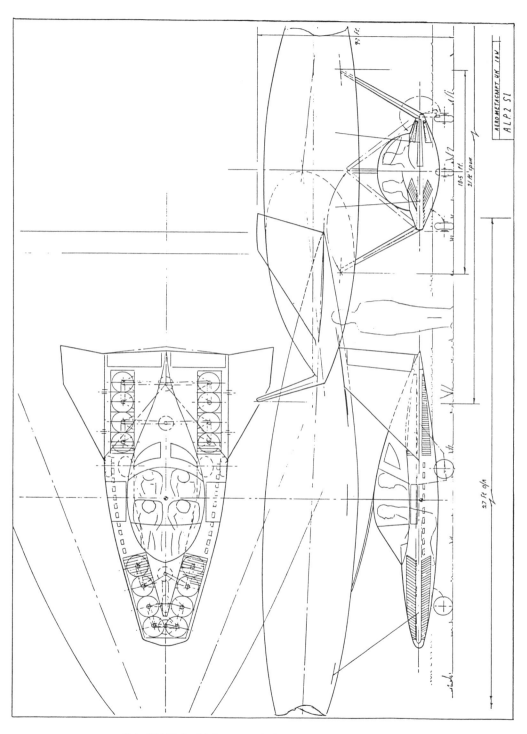

G.A. CMAL 2 with larger tail plane extended as support for
inflatable lifting body type wing.

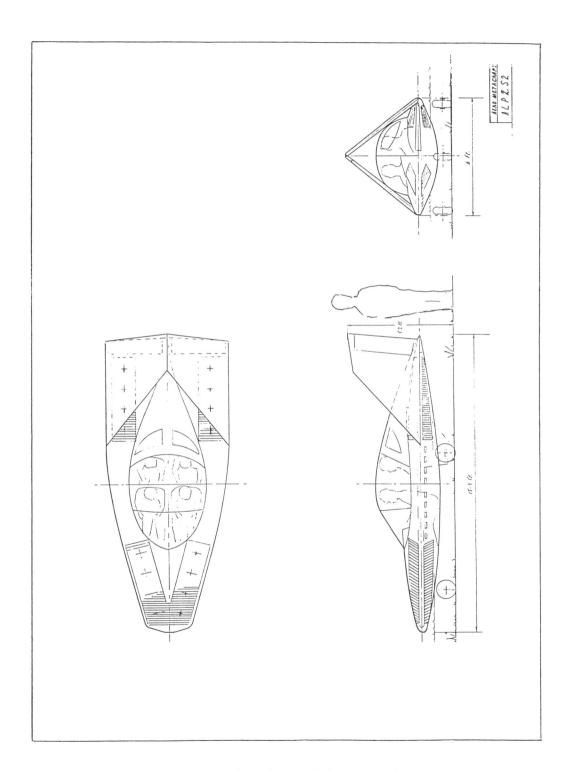

G.A. CMAL 2 with larger tail plane retracted.

1986. The CMAL 3. A high speed VTOL submersible aircraft.

As with the other members of the CMAL family, the CMAL 3 employs a multi-fan small engine system for lift and propulsion. This not only helps to meet low production costs, centre of gravity variables and VTOL surface effect capability, but also due to its small fan diameter system it lends itself admirably to designs for practical fully submersible VTOL aircraft.

This CMAL 3 version is arranged in a lifting-body, tandem seating, canard format which has a greater surface area than the CM 200 canard with corresponding maximum surface effect lift, while the performance/speed range is somewhat lower than the CM 200. Thereby within a cost utility size bracket, there is a fairly comprehensive range from which to choose.

*CMAL 3 high speed
VTOL canard in road,
hover or surface
effect modes.*

*CMAL 3 in forward
flight or surface
effect modes. Note,
in all surface effect
modes U/C can be
retracted.*

CMAL 3 showing canard configuration and lifting body fuselage. In sub-aqua mode, over 80% of the volume is flooded.

Wings folded in water-borne or surface effect modes.

Wings extended preparatory to dive sub-aqua mode.

CMAL 3 in sub-aqua mode in which the normal aerodynamic control surfaces become marine environmentally functional.

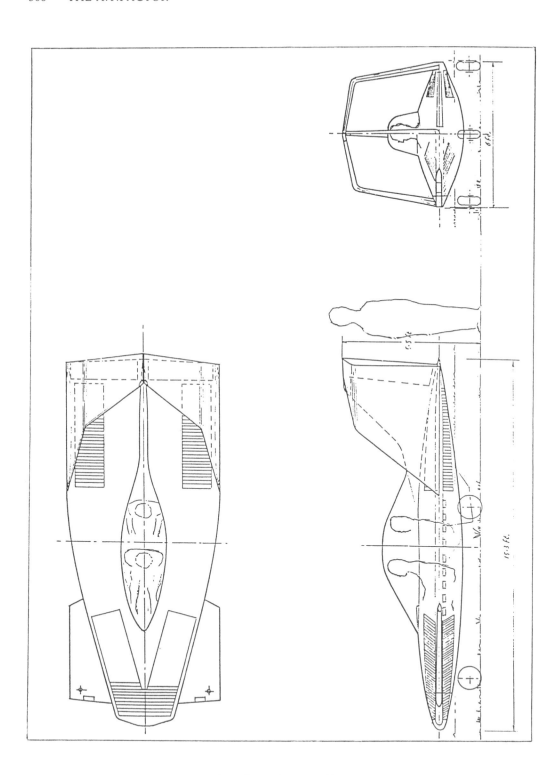

G.A. CMAL 3 wings folded.

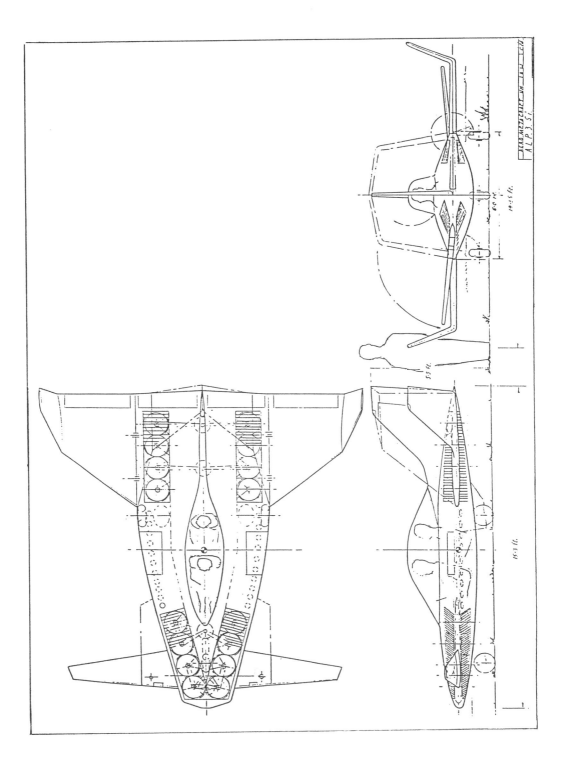

G.A. CMAL 3 wings extended.

1987 - present. The CMAL 4 Puddle Jumper variant of the series, designed to examine a new propulsion/ lift principle.

This two plus two seat prototype test vehicle was also aimed at a mass market. It employs a special mixed fluid propulsion/lift system developed by the author which embodies some of the basic principles portrayed in the foregoing pages. Regretfully however this cannot be discussed further here. The Puddle Jumper operational sequence will be similar to the CM 200.

Although the foregoing general arrangement drawings shown in this book are accurate in that they have been reproduced from the originals, the essentials, such as structural details , loadings, power, air mass flow, aerodynamics etc., would be out of place here. However the designs represent practical well within the state of the art vehicles and (especially in the role of surface effect/road application) several could lend themselves or kit potential, in much the same manner in which small conventional hovercraft became popular. In addition it may have been assumed that stability and control in the main axes while in hover, will be achieved by centre of thrust shift via a rate sensing system. However the author has anticipated the application of a unique alternative currently being investigated.

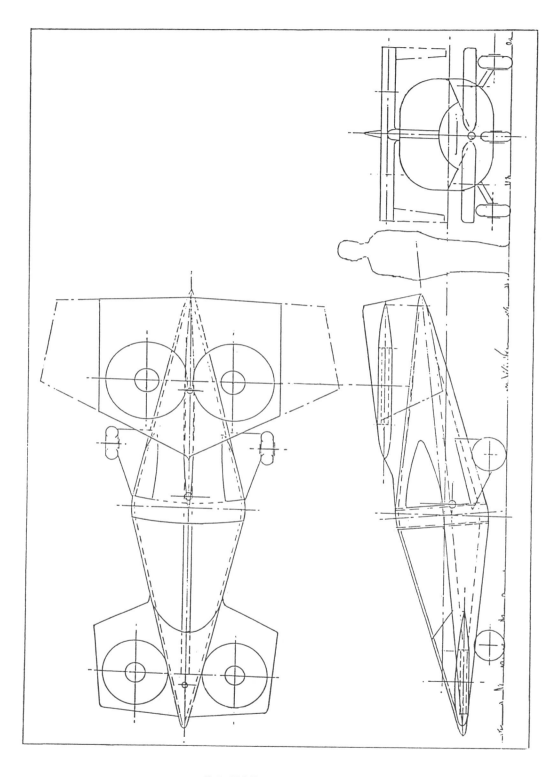

G.A. CMAL 4 Puddle Jumper.

First fit jig assembly of Puddle Jumper's fuselage showing main duct/longerons. Right. Fuselage nearing completion viewed on engine bay awaiting ply skinning.

Puddle Jumper prototype in stages of construction, cockpit transparency omitted. The whole of this fuselage weighed in at a mere 68 lbs, made the arithmetic worth while.

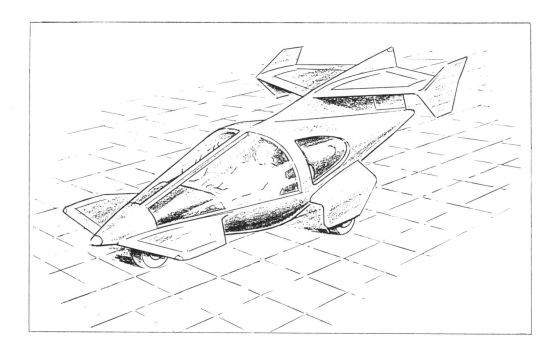

Author's original impression of the CMAL 4 Puddle Jumper.

A threefold closing theme

At first glance, other than by way of a closing theme, it may appear that the subject matter in the following concluding chapter is somewhat divergent from the rest of the book. Therefore it will help to be reminded of the primary objective which has a threefold character viz;

One. To offer first hand evidence which supports some aspects of the (to us) paranormal, including UFOs, precognition and/or time slips.

Two. Whether precognitively engendered or not, how difficult it can be for most of us to initiate and protect new ideas and how irresponsible some members of our society can be in this respect.

Three. Not least, this work is an attempt to correlate and offer evidence for the author's credibility in making the extremely important declarations concerning a future transportation and energy system presented in the recent book The Cosmic Matrix.

PART FOUR

Precognitive Excursions
of another kind

12
Anti-dazzle to Anti-gravity

1958 - 1992 An anti-dazzle device with precognitive overtones. ATF 34 years.

It has been said that on many occasions the overlapping of these various projects has paid dividends, the anti dazzle device was just such a case. Initially this may appear to have little to do with aviation gravity or energy, but in point of fact it embraces three other contributions. First the idea that a certain line of investigation can act as a catalyst for another completely different application is illustrated by remembering the occasion when in the 1950s the author's brother referring to the `Hoverer' model said, " Wouldn't that make a smashing lawn mower".

In this context it should be made clear I have no illusions about the emotional association of inventing something, it's a good feeling of something achieved. However most engineers are aware of the fact that there never was, nor can be, the one all embracing elegant solution to an engineering problem. There has to be areas of compromise, one hundred per cent excellence in all areas; manufacture, economics, performance etc., can never realistically be achieved, therefore one inventor's solution to a specific problem might be better in some areas than another's. One tries to balance these mindful of the fact that in some aspects your opposite number may produce something, which for some reason or another, will prove to be a more acceptable proposition. This is a fact of life we can and do accept.

However the second associated contribution may help to highlight the fact that good idea or not, some people have an inexplicable reluctance to even peripherally consider a new idea and furthermore go out of their way to find justifiable reasons for their behaviour. The inventor has to hope he finds the right ones!

The above idea came to the author as such an off shoot while experimenting with electromagnetic, electrostatic and gravitic effects in which polaroid screens were used, together with an amalgam of constantly being dazzled by car headlights at night. The accompanying rough sketches portray the basic principle. The only penalty incurred would have been slightly more powerful headlamp bulbs, while the plastic hardware costs would have been minimal.

On recommendation a report was prepared and duly submitted to the then NRDC (National Research and Development Corporation) for consideration and/or comment. Almost by return of post I received a letter turning down the proposal viz, `Although the idea is considered technically sound, it would nonetheless be extremely difficult to implement due to its successful adaptation presenting more of a social rather than a physical problem'. In other words, unless fitted to every vehicle by an act of law, the alternative would require every car owner to stand the cost of fitting this system to *their* car which would be quite useless unless *other car owners did the same.* Thus we would all be paying for a unit which would benefit someone else and vice versa.

Now this is obviously true, but in view of the current legislation requiring the use of seat belts....It does have awesome overtones when one considers the number of lives and damaged bodies which may have been caused by headlight dazzle on the world's roads at night since 1958!

Finally the third associated contribution may not be immediately obvious, yet its implication may turn out to be the most important of them all. For within the context of the predominant pattern in this book the reader may recognize an underlying thread among the foregoing precognitive examples which does rather suggest there *is a right time and place for everything.*

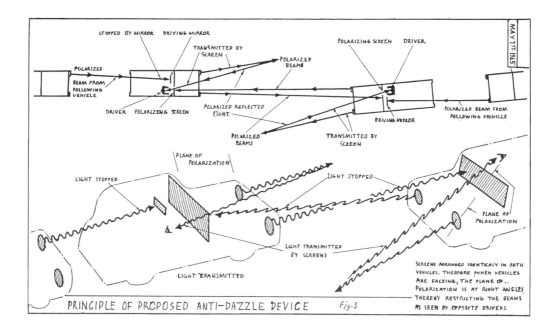

Copy of original sketch submitted to the National Research and Development Corporation in 1959.

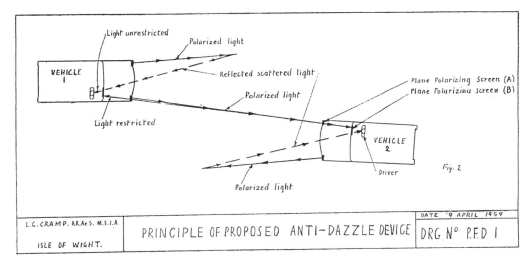

Copy of provisional specifications drawing 1965.

Mail on Saturday

MOTORING

Right on the beam!

Saab's new headlights help you see twice as far, and don't dazzle

Daily Mail, Saturday
September 12, 1992
PAGE 50

by MICHAEL KEMP

REVOLUTIONARY dazzle-free headlights will soon allow drivers to see twice as far at night in busy traffic. Even when dipped, the headlights will reach objects up to 150 yards away, without causing a hint of dazzle to on-coming drivers.

Incredibly, the headlights will be invisible to them. Unless it is lit by street lights, the road will appear to be totally dark. The secret lies in ultraviolet light, which cannot be seen by the naked eye. But the objects that UV light illuminates can be seen.

Florescent paint, many articles of clothing, jeans, bicycles, parts of vehicles, road markings and reflective signs can be seen clearly, as outlines or glowing brightly. The light also pierces fog.

Following extensive development in Sweden to find an answer to the growing danger of inadequate dipped beams at night, Saab says it aims to introduce cars with UV headlights within two years. The cars would then need florescent striped fronts and backs, in conjunction with side lights, to be sure of being seen — and a change in the law may be required.

Saab, owned by America's General Motors, has volunteered to share its research with BMW, Mercedes-Benz, Volkswagen, Porsche, Vauxhall, Rolls-Royce, Jaguar, Fiat, Lancia, Alfa Romeo, Ford, Volvo, Renault and Peugeot — at no cost.

The firm says that if UV lights were adopted universally, night dazzle would gradually disappear as older cars were scrapped.

Traffic congestion is now so bad that 97 per cent of night driving has to be on dipped beams, which light just 60-75 metres ahead. That's a safe distance for speeds of no more than 30-35mph, says Saab, yet on motorways much dipped-beam driving is at 70-80mph.

Main-beam lights sometimes cannot be used, even when they are necessary, for fear of dazzling other drivers.

In Sweden, Saab has already persuaded the government to paint sections of road

with florescent markings to aid research and development.

In 1994, Saab plans to produce cars with low intensity 15-watt UV bulbs that will give a normal light of 50-watt strength on full beams, but will change to ultraviolet on dipped beam. Dipping the lights will drop a filter in front of each bulb to cut out visible light, allowing UV only.

UV light passes right through wheel spray, mist, fog, rain and even snow.

Saab is solving the potential problem of not knowing whether the lights are on by wiring in solar and action sensors to prevent use of the headlights until it is dark and the car are moving.

Saab says the lights will cost little more than current headlights.

For 1993, Saab is introducing an engine management computer which keeps exhaust fumes cleaner than air in many traffic-congested streets.

Called Trionic, it comes with the latest Saab 9000 2.3-litre turbo models at no extra cost. The cars cost £24,895-£28,350.

This article in the Mail on Saturday February 12th 1992 (over three decades after the author submitted his proposal) is one more example of inherent inertia, or how difficult it is to energise people into getting things done.

1950. The EM Hydro-drive. The author's proposal for a hydro electromagnetically propelled submarine craft. This same principle currently investigated in Japan was shown on the BBC Tomorrow's World program in 1992 as a `new advanced concept'. ATF 42 years.

In the enclosed extracts from my articles it will be seen that I anticipated the use of the induced flow of the ejector principle - both for cooling the coil and increased thrust - as used on the cold turbine engine shown earlier, but so far not mentioned in the Japanese project.

However, due to the nature of the electromagnetic principles employed in this concept, it is deemed necessary to emphasize that it has no other relevance to the final conclusions shown in The Cosmic Matrix.

Today of course this same principle is recognized as MHD (Magneto Hydro Dynamics) and it is interesting to note that despite the fact that it depends entirely on a reaction mass, either fluid or gas, and would therefore be quite useless as a space drive - and to say nothing about the extremely erratic high g manoeuvres displayed by UFOs, which would normally tear the vehicle's structure and its occupants apart - not surprisingly it is only non Ufologists or non technical people who give MHD as a main drive a second look in this respect.

ELECTROMAGNETICALLY PROPELLED MARINE CRAFT

As a matter of interest to those who are unfamiliar with the phenomena a little explanation may be welcome.Briefly it is brought about by an induced current effect identical with an ordinary transformer. That is, alternating current supplied to the coil of an electro-magnet produces alternating electro-magnetic lines of force continually cutting the ring (or conductor) in which a current is thereby induced. This current creates a magnetic field which is always of opposite polarity to the field of the alternating electro-magnet inducing the current, hence the repulsion and the heat.

Some early experiments were carried out on model submarines judicious use being made of soldered up empty tin cans for buoyancy tests in a garden rainwater butt!

The accompanying extracts from the author's articles show how this concept was a direct result of experiments in alternating current induced field lift discussed in the books 'Space,Gravity and the Flying Saucer' and 'Piece for a Jigsaw'.

ELECTROMAGNETICALLY PROPELLED MARINE CRAFT

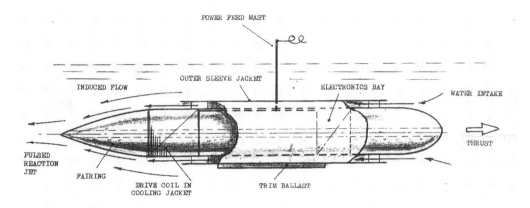

Schematic Layout of Water Jet Thrust Unit.

The Author reasoned that salt water (acting as a
conductor) should similarly be repelled from a
specially designed induction coil immersed in
water, thereby producing a pulsed reaction jet of
useful efficiency. The accompanying simplified
sketch portrays the basic idea.

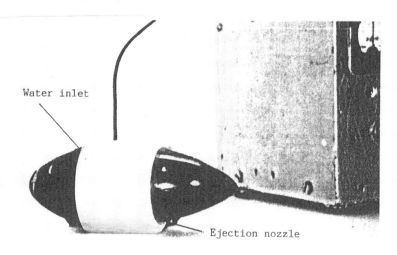

Experimental Model.

Water tank test rig 1950.

1962 - 1971. First hand examples of intrigue in the toy trade give an indication of the situation which would have to be met over a potential world wide technological `breakthrough' involving a transport/energy system.

Miniaturized engineering

During the many years of my association with aircraft - in particular VTOL machines - I became familiar with and intrigued by the principle of the ducted fan as demonstrated by Dr. Alexander Lippisch (later coined by engineers as the `Lippisch duct). The merits for employing this arrangement, together with its disadvantages, are more fully dealt with in most aviation text books, Briefly however it is well known that the thrust of a fan or propeller can theoretically be almost doubled if it is placed within a large bell-mouth shaped duct. However designers are well aware of the fact that when such a duct is tilted at an angle of attack to the oncoming air stream, an aerodynamic lifting component can be obtained from the upper and diametrically opposite lower lips much the same as in biplane configurations) .

The efficiency of small model aircraft used for testing ideas suffers considerably due to scale effects, so that particularly where VTOL is concerned, the use of the Lippisch duct is even more attractive. Therefore quite a few of our free flight models were fitted with these.

As discussed earlier in reference to the Hoverplane, when the models were tested on a rotary beam, the duct could be swivelled about its horizontal axis, so that we could measure the amount of direct *vertical* thrust, or by tilting it slightly out of the vertical we could obtain vectored thrust for forward flight by exactly the same principle as aircraft like the Harrier jump jet. By tilting the duct even more, to say over forty five degrees, the forward thrust component rotated the support beam about its central vertical axis causing the duct to generate `wing' lift.

The first captive rocket toy

Always being tempted to give rein to the aeromodeller in me and build a toy for my sons (plus the ever present need to increase my limited earnings) I decided I could profitably design and build a make-believe rocket model which employed this ducted principle. It proved to be extraordinarily successful. The fan noise issuing from the tail jet was enhanced by the simple expedient of placing a thin sliver of plastic into the fan blades which produced a turbine like note. It was made visibly convincing by placing thin foil streamers in the slip stream which were illuminated by a small bulb within the tail cone which dimmed in unison with the controlled speed of the fan. With the model secured horizontally (as if in an engine test bed situation) truly, some grown men were

The toy rocket in phase one vertical take -off mode with wings retracted and VTOL legs extended, showing augmented thrust fan intakes. The illuminated exhaust cone lies at the base of this.

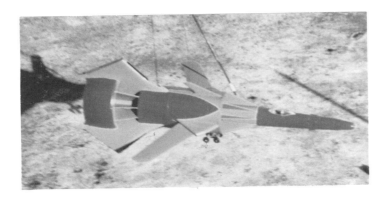

Phase two. Toy Rocket with wings and undercarriage extended for conventional flight mode, VTOL legs folded. These two old photos are all that remained to show for 14 months work.

loath to put their fingers near to what really looked like a hot flame! The general style of this toy `rocket ship' was after the fashion of the `Thunderbirds' TV series, complete with retracting space craft type landing legs, as shown in the accompanying photograph.

The model was both suspended and activated by short lengths of thin transparent fishing line which could be reeled in or out. The whole being attached to a rotary counter balanced beam pivoted at the top of a central `space station' type control tower, also shown here.

Just like the real thing

The operational sequence being, **Phase one.** The model, standing vertically in its VTOL mode on an illuminated scale launch pad was motorized to very slowly proceed outwards to the `launch area' situated to one side of the tower. When the motor was `lit up' the counter balance rocket would very realistically take off and hover, during which the launch pad was withdrawn back to the tower. **Phase two.** Had three simultaneous functions viz. The swing-wings began to extend (Tornado style) together with retract of the VTOL landing legs and vectoring tilt of the whole model to conventional flight mode, which initially produced a slight rotation of the support beam, then increasing rotary speed to maximum full ahead, *which could be very dramatic !* In this way the `pilot' could both slow the thrust fan and dim the `exhaust' luminosity together with increase or decrease of flying angles. Thus in this mode the model could either climb, or if desired, descend to a perfect Concord type landing.

Enormous play value

Alternatively the whole procedure could be reversed, with conventional take-off and several jumbo jet style circuits, to full speed ahead climb, then slowly rotate to the hovering mode. The launch pad was then trundled out to the touch down area and the skill was to position the model, now with wings retracted and legs extended, to perfect VTOL mode touch down on to the pad. We had adults - including pilots - visit our home who were quite unable to resist playing with this toy. Not least myself !

The design allowed for all kinds of `extras' ranging from different formatted vehicles, moonscape, rotating raydomes etc.,etc. Not least we made the toy so versatile, it could be used by a youngster in a hospital bed. I successfully took steps to produce a cost effective toy by employing all sorts of simple ideas allowing the basic unit to be folded up and housed within a modest size box. However with the ever increasing extras for this already over sized prototype, my transport box was five feet long, eight inches deep and fourteen inches wide, not one would think, a very easy thing to overlook or lose, yet that is exactly what is supposed to have happened.

Having worked sporadically on the model over several years, I had ironed out all the bugs and got it to pre mass production stage. Then I approached the big toy companies, the first being Mettoy Ltd., UK, whom I visited in company with my wife as demonstration assistant. The meeting was traumatic, the managing director Mr.Henry Ullman and his colleague acted as if completely mesmerized. Promptly took us to lunch and offered me, my wife and son, a permanent R&D design job on the spot! Yes, they would produce the rocket toy subject to the usual several manufacturing requirement changes and drew up an attractive contract for royalties to take home and consider.

Several weeks and meetings later found me in a tight spot. The company were putting pressure on me to resign my position at BHC to join them as they felt there was a market for the type of `space age' ideas we had, but they required our specialized knowledge to develop them. Unfortunately the situation reached a ` we will only go ahead with the

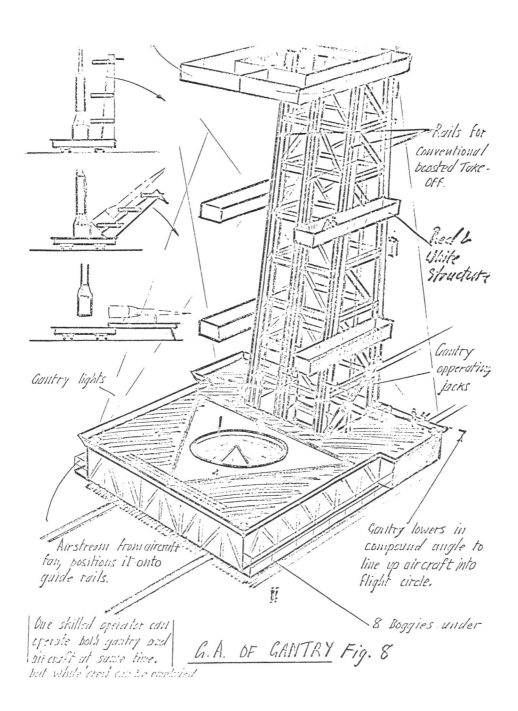

Rails for Conventional boosted Take-OFF.

Red & White Structure

Gantry opperating jacks

Gantry lights

Gantry lowers in compound angle to line up aircraft into flight circle.

Airstream from aircraft fan, positions it onto guide rails.

One skilled operator can operate both gantry and aircraft at same time. but whole crew can be employed.

8 Boggies under

G.A. OF GANTRY Fig. 8

One of the original sketches for the Toy Rocket Plane.

rocket toy if *you* do it for us' situation. This was an unexpected and difficult spot to be placed in, but fortunately I had taken the precaution of contacting two other large companies in case they might offer me an option. The first of these had been the Lesney Matchbox Toy Company of London. They also received the toy with much enthusiasm, but at that time they were heavily committed and had an ongoing problem involving the large American company (Mattel Inc., California) which was seriously curtailing their R&D program.

Apparently at that time Mattel were sweeping the world market with their well known Supercharger type toy motor racing track system, in which small freewheeling matchbox type cars were ejected at high speed along banked plastic tracks. Lesneys didn't beat about the bush and said that although they couldn't catch up with Mattel with their product - for which Mattel held the patent rights - they had to put something similar - but slightly different - on the market as an `image saver'. Even though it might entail a complete financial loss for them, could we come up with some ideas `to get round the patent'- but *very* quickly!

By now I had a fair idea about the world of patents being a very suspect one, but it came as something of a shock to be asked to participate in something which seemed to us a

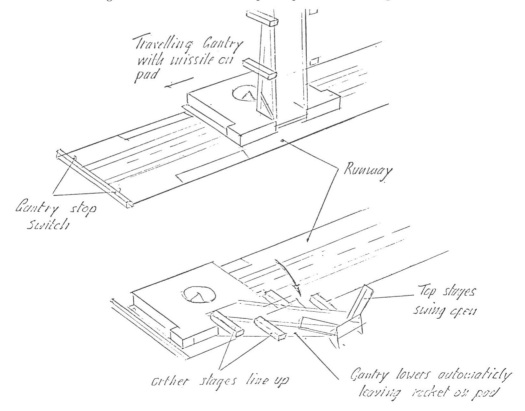

Another sketch depicting the Rocket Toy launch pad and runway traverse.

dirty tricks game. I said I would think it over but determined that if we offered something *it would be different* and fairly competitive.

What price patents?

In a very short time Gary and I produced no less than five ideas, all of which *were* not only completely different but cost effective and superior in performance to the Mattel toy. We were duly paid for the long days (and nights) spent, yet despite it all we were amazed to discover that in order to save the new tooling costs that our solutions would entail, Lesneys had gone ahead and produced what we thought was an almost carbon copy of the Mattel toy! Later when I enquired,`what about the patents?', the matter was casually shrugged off. However we accepted Lesneys offer to work freelance for them in other R&D work for which we received a retainer (sometimes barely sufficient to cover the material costs), but it was fascinating work and we profited in other ways from this brush with the toy trade. We were promised royalties for any successful product we submitted, but despite the fact that there were such, the royalties never materialized. Of course I had informed Lesneys of the affair with Mettoy (which by now had fallen through due in part to a bad year on the market together with our steadfast position). Meanwhile I had received an invitation to demonstrate the rocket toy to the UK branch of Mattel in Northampton. They also showed great interest and asked if I would be prepared to have it shipped to their head office in the US, to which, after due consideration I agreed.

After the first few weeks of silence I enquired by phone as to the position. The manager was unavailable and the secretary didn't know. Then the letters began, more and more phone calls, evasive answers, different works managers came and went back to the US etc., etc. This frustrating situation went on for the best part of a year. Nobody it seemed knew anything about my five foot long package! Finally in desperation I had to make private enquiries which produced the name of the transport company involved. Yes, they did remember the package and produced not only the number, but the date and the name of the truck driver who had collected and *delivered* it to Mattel Inc., California.

Armed with this information I wrote again to the UK branch who answered to the effect that they were just as mystified as I was, they were most apologetic, but couldn't explain where my king size package had gone. However they said they were most concerned and I would be hearing from them.

Some weeks later I received a letter from a representative of Mattel who said he was in the UK and would like to visit me on the Island and suggested a date and a venue where we could have dinner. I was mystified but after all this time I was relieved to think Mattel had recovered the toy and the rep might be bringing it with him. However the night of the meeting found me having dinner with a very pleasant gentleman who seemed genuinely concerned that in fact he had not brought the toy and as his company accepted responsibility for its loss, he had been instructed to offer me compensation!

I knew then of course that I wasn't likely to get the toy back. It was a hard blow for I had a good idea of its royalty worth which could have been considerable. Towards the end of our meeting the representative said he had in fact been instructed to offer a cash settlement which he was adamant was non negotiable. Perhaps it was a good ploy and sometimes in a more cynical frame of mind I have since reasoned that the long silence, inconvenience, lack of funds and final disappointing acceptance that one has irretrievably lost something, can be an effective softener for an easy touch. So when unexpectedly there comes a crisp cheque (whose value is nowhere like meaningful recompense) it can seem *very* acceptable at the time.

III/5/12/71

DEED OF RELEASE

This Deed of Release is made this 15th day of September, 1971 by and between LEONARD CRAMP of *[handwritten address]* (hereinafter called "CRAMP"), party of the first part, and MATTEL INC., a Delaware corporation (hereinafter called "MATTEL INC.") and MATTEL LIMITED, a company organized and existing with limited liability under the laws of England with its registered office at Rixon Road, Wellingborough, Northamptonshire NN8 4LL, England, (hereinafter called "MATTEL LIMITED"), parties of the second part.

WHEREAS in April of 1970 CRAMP delivered to MATTEL LIMITED for shipment to and inspection by MATTEL INC., a model and sketches of a pylon from which toy planes, rockets and helicopters could be launched and flown by means of an electric current flowing through a series of support arms (which model and sketches are hereinafter called the "model"), copies of the sketches being attached hereto and marked as Exhibit A hereto;

WHEREAS during the course of shipment to MATTEL INC., the model was lost and has not been recovered; and

III/5/12/71   - 2 -

WHEREAS MATTEL INC. and MATTEL LIMITED desire to obtain a release from CRAMP for any and all claims arising out of the loss or use of the model.

NOW THEREFORE, for and in consideration of the sum of one thousand pounds sterling (£1,000) to CRAMP in hand paid by MATTEL INC. and MATTEL LIMITED, the receipt whereof he hereby acknowledges, CRAMP hereby remises, releases, and forever discharges MATTEL INC. and MATTEL LIMITED and their agents, employees and assigns from all manner of actions, causes of action, suits, liabilities, covenants, agreements, damages, claims and demands whatsoever arising out of the loss or use of the model or any invention based thereon by MATTEL INC., MATTEL LIMITED or any other person, firm or corporation which against MATTEL INC. or MATTEL LIMITED CRAMP ever had, now has or which he or his heirs, executors or administrators hereafter can, shall or may have.

IN WITNESS WHEREOF, LEONARD CRAMP hereunto sets his hand and seal the day and year first above written.

Signed, sealed, and delivered by the said Leonard Cramp in the presence of:

[signatures]

LEONARD CRAMP

The Deed of Release offered to the author for the loss of the Rocket Toy in 1971.

The name of the game

About a month later Gary and I entered the R&D department at Lesneys and were met by the marketing manager and his assistant who by now knew of the continuing saga with the lost toy. In silence they unpacked a box and set up a recently marketed toy by Mattel which was merely a model helicopter suspended on a counter balanced articulated beam on a central stand. The marketing manager looked at me and said "Here is a possible answer to your lost toy". He went on "Mattel would have been well ahead with tooling for this little helicopter at the time your rocket toy was delivered to them. It could have represented a *very* expensive hiccup for them. For they could hardly have revealed their plans to you, neither would they have wanted you to take your toy to somebody else". Now I had no way of being certain that this impasse was a reason for the loss of the toy, it may after all have been stolen, but it gave me an insight into the kind of dilemma which could arise when freelance outsiders proffer new ideas around to the establishment. Moreover wasn't it Lesneys themselves who had displayed indifference over the Supercharger? So they of all people knew the name of the game, but of no less importance to me I was being introduced first hand to a system which most of us are only able to guess at from the outside. As shown earlier, this would not be the first nor the last time that some of the author's work would mysteriously disappear. Shades of the British Book Co., New York and Space,Gravity and the Flying Saucer!

What price for toy
ideas galore

During the four years I worked for Lesneys as a freelance consultant/designer, my sons Gary, David and I took part in many extraordinary adventures, save for a sampling few, too many to list here. Among these was a small blowpipe ejected helicopter type rotor which in operation was identical to native's jungle blowpipes. A simple transparent plastic tube fitted with a special fail safe mouthpiece, was end loaded with the small rotor of about five inches in diameter secured to a pre water ballasted central hub. All credit to Gary he ingeniously formed the wings from plain strips of thin flat plastic folded in the fashion of a V in such a way that when ejected from the pipe at speed they spun and centrifugally deployed with a naturally formed angle of attack! In the ascent these small light weight 'missiles' - having a relatively low drag coefficient - could attain remarkable heights, whereupon they would descend like sycamore seeds slowly and gracefully as glistening gossamer with all the colours of the rainbow.

At an ensuing demonstration to the board of directors of Lesneys we were delighted to offer an on site treat in the quadrangle outside the main building and even more delighted when the small boy motivation among them was found to be too difficult to resist!

Then there was the 'Chunnel' which was conceived long before the channel tunnel was commenced, In conjunction with Lesneys own R&D engineers, we worked on it for almost a year. This idea was largely born of my experience with hovercraft and was stimulated by acting on a whim, I placed a small polystyrene ball into a ring made of one inch diameter transparent tube measuring approximately three feet across. The inside face of this was perforated diametrically with a series of small holes, into one set of which I directed the outlet end of my wife's vacuum cleaner. I well remember the effect was quite dramatic!

I next constructed a series of light weight cylindrical railway type coaches fitted with a set of Lesneys 'hot rod' matchbox series wheels, together with small magnetic blobs for couplings. This provided articulation around sharp curves. The 'track' was formed from the same one inch diameter transparent tubes with both straight and curved sections. At high speed, low centre of gravity and centrifugal force caused the carriages to bank and -

depending on the track layout - in some cases run upside down!

Again, drawing on my aviation experience, I built and experimented with suitable scroll type centrifugal fans, similar to, but more powerful than common hair dryers. The most suitable one of these was incorporated into the aerodynamics of the system which was suitably disguised in the form of a railway station terminus. The transparent tubular tracks were divided up to form a two-way system divided into convenient lengths with airtight plug-in end fittings. This layout was complete with track overhead supports and special sidings etc., while the speed control was obtained by the simple expedient of fitting an ordinary butterfly type balancing valve. We found that with this it was possible to control the vehicle's speed from a blurred maximum to the slowest real life start from rest you have ever seen. One of the highlights of the toy was the fact that unlike 00 gauge train sets, the `chunnel' could be arranged over and under room furniture, underneath carpets, into and out of a heap of sand and not least under the water of a garden pond without a spark of electricity to worry about. In fact there was so much going for this toy in terms of both play value and commercial viability, it seemed most unlikely that Lesneys wouldn't take it on. Yet sadly it was shelved. Today we have only a few lengths of transparent track and some little coaches left to remind us of our interpretation of the `chunnel'.

Among the long list of toy designs was a tiny scale *working* helicopter measuring barely three inches in length which - without attaching wires or supports of any kind - could take off by aerodynamic means, hover, move in all three axes and with spinning rotor (and a little skill) be flown completely through a standing wire hoop and pick up tiny buckets and other `payloads'. All of which could be done over a pretty green baize landing field measuring some two feet by three feet, which could be rolled up and the whole contained in an ordinary small size box! The technical know how was simple, the manufacturing costs were extremely competitive, everybody wanted one except Lesneys!

Still with aerodynamics, the `Whirligig' was a wind driven land/water boat, but this one was motivated not by sail but a windmill type rotor geared directly to provide traction via a water propeller or road wheels. Complete with swivelling weather cock which provided continuous wind heading, this little machine could drive *directly* into wind. In 1957 we even seriously considered a full size version! Like its predecessors, its play value was enormous, its production costs minimal, but in Britain..... the same sad story!

For Scalextric type model race cars I offered Lesneys a remote steering system in which each vehicle was fitted with two smaller, rather than the normal one large, electric drive motor, one mounted forward and the other at the rear. Each motor drove only one wheel leaving the opposite wheel freewheeling. This unit was pivoted in the middle with a governed degree of travel backward and forward, so that when placed on a surface and powered, the driven wheel tended to turn the axle about this central pivot and thus turn the vehicle in that direction. By placing the second unit with the driven wheel on the same side as the other unit and by interconnecting both axles by a suitable diagonal link, the following motions could be obtained. Both fore and aft motors powered, turning couples were cancelled out by the link and the vehicle proceeds dead ahead. By powering say only the forward motor, would cause the vehicle to turn in the appropriate direction. By disengaging this drive and powering the rear motor the opposite turn was achieved with varying degrees of full and partial lock.

The proposed technique for electrically activating this system is rather more obtuse to describe here, suffice to add that at suitable track junctions there were to be placed a series of magnetic signal switches. As with the other designs Lesneys claimed they had patented this idea, but again to my knowledge it was never used.

Back to the future

As will be seen later it was during my sojourn with Lesneys that I produced a toy in which a revolving miniature car was rendered invisible. I was able to produce this illusion in two stages, in which the outer body of the small car became transparent first, leaving the internals - chassis, engine, suspension, wheels etc., - clearly discernable, before finally these also completely vanished. Of course this was rendered more difficult by the fact that the viewer was able to look all round the *revolving* exhibit!

I often remembered having seen this effect somewhere many years ago in my boyhood, but couldn't remember the details, save that the exhibit was a radio set in a shop window in London. Then many years later while working for Lesneys, one evening I happened to look out the uncurtained workshop window and saw my reflection apparently sitting at the wheel of my parked car outside! The effect was very real, but was only made possible due to the fact that the distance, height and posture I happened to be in, in relation to the window, was exactly the distance, height and required orientation of the parked car outside. But of no less importance was the fact that the lighting also was absolutely right, subdued room lighting with brightly lit anglepoise bench lamp locally illuminating me and the car outside. Within an hour I was able to proudly display a partly opened full box of matches and much to the consternation of my family, cause the matches within to disappear. I was to learn much later the trick was as old as the hills, but after all these intervening years I had jolly good satisfaction in achieving it.

During the years I have dealt with the toy industry and have seen many questionable examples involving ethics accepted as the norm. From toys to cars and full size aircraft, regretfully the whole gamut of the world's industry sometimes have to adopt the same stance. It is one of the many reasons which has qualified and motivated me to take the decision to freely offer the information introduced in my book The Cosmic Matrix.

One late afternoon in 1991 I had come to do a rather major bush clearing job in my sister's garden - sadly overgrown - when I suddenly spotted rusty metal. Curious, I dug deeper to reveal whatever it was and the next instant old memories brought a lump to my throat. I looked around at the other houses in what had since become a busy cul-de-sac (shown in Chapter 4) and remembered the time when I last held that rusty old piece of metal. Then the garden had been trim and lawns mowed and there had been only one other house to be seen over one hundred yards away. That was Gardeners Cottage the the house in which I and my family once lived and UFOs had their sway.

At that time in 1956 I was employed in the MDO (main design office)of the then Saunders Roe, working on Britain's first space rocket, the Black Knight. But as was seen in the foregoing extracts from my articles, at that time I was privately researching the behaviour of disc wings.

One of the first things I had established was that centrally balanced discs are totally unstable in flight due to the forward positioning of the centre of lift. But it was also well-known that this can be easily rectified if the disc is rapidly rotated about its central axis, as in the case of the familiar `Discus'. But what isn't so well-known is the fact that although the tendency for a spinning gliding disc to stall is thus negated, the gyroscopic precession produces a rolling motion at right angles to the direction of flight   In fact I had devoted a whole section on this phenomenon in my UFO book Piece for a Jigsaw in 1964.

In 1956 in order to control this tendency I had experimented with such spinning aeroforms and finally obtained success by the simple expedient of forming short stubby aerofoil blades around the perimeter, so shaped as to create asymmetric lift in opposition to the roll.

Thus it was quite a normal sight to see me hurling the discs between my sister's house and my own, also shown in Chapter 4. So much so that when my brother and his young family joined mine for our annual reunion, it was normal procedure for them to ask for a disc to throw around. Experiments had revealed that concave dishes provide more lift and give the flattest glide. Knowing this and despite my long acceptance of the UFO phenomenon (and the fact that I like to keep my feet on the ground) on one occasion in order to illustrate how easy it would be to fake a UFO photograph, the boys and I took one such upturned dish to the highest point of the Downs, attached a lighted mole smoke cartridge to the underside, and hurled it - discus fashion - into the air. From the fake point of view the resulting photograph was highly successful, showing an apparently white disc cascading and streaming a convincing vapour trail! This discus fun went on year after year and we all became quite good at discus throwing. The most successful of all the discs we used was a pressed pan shaped dish commonly named in those days as the Fray Bentos pie tin.

That evening in 1991 as I stood there alone with echoing memories of laughing youngsters, I once again experienced that odd response conjured up by my brother's chance remark at that very same place many years ago, when concerning the `Hoverer' model he had said "Wouldn't that make a smashing lawn mower". Mindful of the déjà vu I glanced down at that rusty old pie dish still with the remnants of the original makers lettering and reflected, in my efforts for aerodynamic supremacy of the disc, I had once again completely overlooked a no less meaningful adaptation for my Fray Bentos pie dish. Today it is called a `Frisby', and made its eventual originators *very comfortable indeed!*

One of the last toys I dreamed up for Lesneys was a building. One day the directors said they had another problem. Unlike their mini Matchbox toy cars - which could easily be housed in conveniently small buildings - their much larger King size series were far too big for marketable storage buildings. Had I any ideas? Again, aided by my experience in hovercraft, I used one of my `Chunnel' fans to inflate a polythene dome, over three feet in diameter and eighteen inches high. It was constructed from several seamed gores, and when deflated it could be packaged into a small box measuring only six inches by nine inches. It was fitted with a series of foldable plastic wedge shaped foundation/sealing sections, which allowed a parachute type roll up technique. Slightly weighted plastic doors gave constant access for the vehicles without noticeable deformation or demand from the occasional moderately running fan.

The `Kingdome' could be illuminated at night and looked great with its surmounted fluttering flag. I gave Lesneys this three foot diameter building for their King size toys, they were delighted and I had much fun building it, but again sadly it too was never used!

In 1975 I joined Desmond Norman's design team at N.D.Norman Aircraft, Isle of Wight, working on the well-known `Firecracker' aircraft. It was during that period, when having discovered my association with the toy trade, Desmond introduced me to his brother Torquil who owned `Timpo Toys' in Scotland. At one time, before a near fatal accident, Torquil had been an internationally well-known skydiver. So perhaps it was natural for them to decide to produce a series of model aircraft which would carry the signature of Desmond as the co-designer of the well-known Britten Norman `Islander' aircraft. Accordingly they invited me to join them as the model designer based on some of Desmond's proposals.

Naturally I was delighted to take this on and with the help of my sons Gary and David, plus the constant help of my wife Irene, we burned the usual midnight oil for several months building twenty completed models, boxed and ready for exhibit at the imminent toy fair in Germany. Desmond had proposed three ideas to be produced with vacuum formed fuselages and foam wings, comprising; 1. A propeller driven fighter/trainer aircraft (not unlike the Firecracker). 2. A fully high density plastic sailplane, and best of all, 3. A Stinson Reliant high wing look alike, equipped with `bomb doors', from which at power run out a pilot with parachute was ejected! No small order to us, for it meant that each model would have to be designed, built from balsa et., and flight tested before I could even begin on the manufacturers machine drawings. It meant a great deal of work and fatigue, but a great deal of fun, topped with an added bonus - that of seeing some of the models displayed in toy shop windows.

It seems pertinent and not without a degree of importance to mention that although the foregoing work often ran concurrently with research of the `more serious kind', it is interesting to note that despite the seemingly poles apart nature between it and the realms of physics and cosmology, there has nonetheless been a recognizable cognizance. From providing the means of acquiring often recycled materials at the mundane level, to discovering remarkably convenient research technological methodology at the theoretical level of the kind revealed in the Cosmic Matrix.

1973 - 1992. The futuristic `Vistarama' for leisure and education. Muddle and intrigue at the council. ATF 19 years.

When I moved home and family to the Isle of Wight in the spring of 1956, I knew from my brother who had already lived there for several years, that the island depended heavily on its tourism and holiday resort attractions which offered mainly seasonal employment for many of its population. Saunders Roe my own place of employment, Plessey Radar and J. Samuel White being the only large industrial companies. Indeed on occasion the Island has been described as a semi-depressed area in need of European support. Therefore as with many others, it was natural for me to reflect on this sad situation now and then.

A useful trade off and a
design for the future

As I have said, one of the better aspects of my association with the toy industry was the pleasure that exciting new ideas can bring. Although I didn't make a good living from it, nonetheless it was a satisfying feeling to see a youngster having fun with a toy glider which I had designed. So I began to form the idea to help provide a place of participating entertainment which would be of pleasurable educational benefit to the young, and relaxing enjoyment for people of all ages, plus a possible attractive income for the island.

First I drew up plans for the type of building which would blend most suitably with the surrounding environment. I visualized it in a park like setting and have halls capable of multiple usage, from games to lectures, and/or libraries etc. The building would be essentially in two parts. One, a large circular domed structure with central hall and a fixed astronomical local terrain telescope remotely controlled with TV screens in a suitable lower room, (reminiscent of the famous Camera Obscura of olden times), and the entire outer circular wall comprising separate walk through rooms with functional exhibits, tea rooms etc.

Two, a large adjacent rectangular hall formed the main participation area, in the centre of which stood a full size realistic `vehicle' equipped with circumferential viewing windows. The interior of this was set out similar to the Star Trek `Enterprise' with the passengers being centrally seated. This `craft' was to be equipped with a wheeled undercarriage mounted on a vertically accelerated continuous belt system. Twin tracks led from the wheels to large doors set into one of the distant end walls which supposedly retracted exposing an equally large, make believe, descending tunnel. Animated illustrated screens in the vehicle showed this layout diagrammatically with the tunnel ending on the sea bed a mile or so out to sea.

*Sectioned sketch of Vistarama main domed hall with central dioramic
TV viewing screen.*

*Copies of original interior layout sketches, the above showing one
circumferential exhibit room.*

By using technology (which today would be similar to that currently employed at Disneyland, but completely unknown then) the circumferentially placed wrap round specially double glazed windows in the vehicle doubled up for multi back-projection purposes. The male and female `crew' and staff members were to be kitted out in uniforms with suitable logo insignia and prior film footage would be made showing them going about their business as it would be on a typical day in the large `hangar'. The format for a given `journey' was as follows.

Having been seated and instructed as to procedure, the passengers would have noticed via the cabin windows, the `flight' preparation team milling about outside. The cabin and outside lights were to be dimmed almost to extinction in order to allow transfer of the real life view outside to be substituted by the back projection process of the forestated similar view previously made on film. Cut into this film was the `journey' section previously made with the use of models, the first part of which showed the continuation of the outside real life event stripped into a shot of the large doors opening displaying the descending tunnel which appeared to get closer as the ship approached it.

At the audio start of the `engines' the continuous belt track was to be motivated thus revolving the undercarriage wheels with slight vertical acceleration in order to convey the feel of motion, coincident with the visual inputs described. Thus the beginning of the journey down and out on to the sea bed was initiated. This was to be standard, but the journeys could be varied in which the vehicle `ascends' vertically from the sea bed showing initial Isle of Wight views as it continued to rise into space for a journey to the moon or such like, aided of course by the visual and motion inputs.

One sequence we developed was an under water exploration. Another took the passengers on a trip around the island back in time when the Romans occupied Britain. Yet another showed how it might look in the future, etc., etc., always with full participation of the passengers. On the moon landing for example it was an easy matter to show a team of `astronauts' advancing towards the `landed' craft carrying samples of `moon rock'. This back projected film sequence would then have been superimposed by live views of the astronauts passing by the windows of the ship, the hangar being blacked out at this time. They were then seen to enter the airlock and would hand out samples of `moon rock' to the children present. We perfected these electronic techniques working on a shoe string budget. They not only worked, but were totally convincing.

We anticipated that once successfully operated, other Vistaramas would be built abroad so that copies of the filmed parts of the journey could be sent to them to save them the expense and/or lack of the necessary know how technology. In all we were totally convinced the project would prove to be a recreational, financial and job providing success for the island.

Accordingly we built small working demonstration models as well as a scale model of the complex shown here, and took them to an appointment with the Chief Planning Officer of the County Council at that time. He was always very impressed and felt the proposal warranted investigation, providing a suitable site could be found and necessary funds raised. I knew this would not be easy due to the fact that the island people are properly environmentally sensitive to developments of this kind.

During the time of developing the idea I had already been meeting business people to try to raise funding, not entirely unsuccessfully. In my efforts I also got the support of the island's MP, at that time Mr. Mark Woodnut, who also helped with directives for some possible government funding. However finding a suitable site in mainland Britain is one thing, finding it on the Isle of Wight is another. We visited site after site only to be turned down, if not by the landowner then by the Council's Planning committee.

Original plan view of the author's Vistarama showing the two main buildings.

Mock aerial view of a scale model Vistarama complete with Lesney matchbox toy cars and a Saunders Roe Black Knight rocket. A redundant full size rocket shell had been tentatively offered to the author for this use.

Scale model Vistarama showing general approach view.

*View with dome removed showing upper, inner and outer promenade
viewing decks.*

A council's alternative
& a council's bungling

Then after many disappointing months the Chief Planning Officer asked me to revisit them. It seemed a site had become available, but it would be too small for my Vistarama project as it stood. However they suggested it might be big enough for part of the general scheme, but run more on the lines of a museum with some of my functioning exhibits.

In a way this situation was very similar to my previous experience with Mattel in that I had lost out on the main course, but at least I was being offered a small piece of cake. With much the same philosophical reaction I was glad to accept this offer and promptly got down to restructuring some of the ideas this entailed. The site chosen was at Fort Victoria near Yarmouth, one of several partly demolished forts around the island. This entailed drawings of the remaining buildings and monitoring traffic flows on the approach road (which for months Irene and I did for the council every Sunday evening, using a traffic recording meter) We were offered one unit in which to install a restaurant and toilets, and although this was to be a mere shadow of Vistarama, the council were prepared to give us two years grace. Accordingly I left my job at the British Hovercraft Corporation, sold my house and took out a loan with the bank to develop the idea myself. In all Irene and I worked for over a year. As already discussed in Chapter 5, the new house we took over had one asset that I would require in which to construct some of the functionary displays, the large barn type building.

Then one day while visiting the council to check on the boundaries of our new property, I quite coincidentally met on the stairs, the council's legal officer with whom I had frequently liased. As I hadn't heard from them for some time I enquired about the outcome of a pending case involving wayleave right of way for the power cable to the museum and restaurant buildings. True there were other people with him so I had to be brief, but I could not believe what I heard. Rather red in the face with understandable embarrassment, he said that at the last council meeting the chairman had *overruled my project* due to the site `not being surplus to the council's requirements'. In other words members of various departments of the council, including the legal officer himself, my wife and I, had all been working under the assumption that this first step had already been taken several years ago! The solicitor said he thought it would all work out OK but it could take months! Needless to say I was not impressed even more so by the almost indifferent attitudes of some of the other council members when I later contacted them. I'm afraid they certainly displayed something of an `I'm all right Jack' attitude.

Perhaps this fiasco was prophetic because I was now forced to resume work in aviation which fortunately I still had. However nine months later, while we were paying a visit to the site I noticed that despite the still imposed wayleave clause, the electricity company had already laid the power cable, together with water to *our* restaurant site, moreover there was (albeit temporary) a cafe complete with owner installed, with never a word from my friends at the council!

In the late afternoon gloom of that autumn day Irene and I walked around nonplussed, we saw the wooden seats and chain posted no-go areas complete with barbecue which was supposedly going to be *our* business and I couldn't help feeling sad and sorry for myself, that is until I noticed one freshly varnished chain post already cleaved by vandals and I found myself thinking `well of course all that will be covered by insurance'. Then I remembered *just one* of my proposed intricate exhibits, (based on the same technique I had developed for Lesneys) of a large scale model showing the fort as it used to be - complete with soldiers on parade in the square, gun ramparts and tanks - which by a press of a button a youngster could make slowly disappear leaving the ruins of the fort as it is now, complete with holiday makers and cars in the car park, where moments ago stood serried ranks of troops. I stood there somberly reflecting, with such vandalizing prospects no amount of insurance settlement could ever induce me to go on building such things again and again.

Today the restaurant *has* been built, as have the toilets. Missing hand rails on the flat roof have been replaced so that many people can now enjoy the lovely views from there (which I had previously suggested and had been frowned on by some council members). There is an aquarium and exhibition room, outside tables and chairs, the barbecues are popular and the car park quite full. My site drawings have been used and filed away and not least, by a strange ironical coincidence, as I pen these words, there has been an application made for the use of one of the rooms for a Planetarium, which has since been installed and is a great success. As for us, Irene and I never heard another word from the council!

Lest some readers are tempted to misinterpret my motives for publicly airing such short comings in society I say be patient, for it is such human frailties as these which have to be carefully weighed up when offering a possible premature inheritance of powers previously beyond the lay public's dreams, as that set out in the Cosmic Matrix.

In the process of developing Vistarama, several of the author's colleagues at the then Saunders Roe establishment kindly rallied round to offer support, among them the late Dick Stanton Jones (who was involved in the visiting Generals episode related in Chapter 3), Tom Pattinson a close friend, physicist at the Black Knight rocket test facility at Highdown, Isle of Wight, and Ray Wheeler who retired in 1991 as Systems Support Director of Westland Aerospace Ltd., who was instrumental in obtaining the Black Knight rocket table top display model for me, which for many years, dusty and worn, used to adorn our MDO. Somewhat selfishly I have since been reluctant to part with this old model which with its unrequited memories, still resides in pride of place in my workshop. Optimistically perhaps, I still like to think its silent inspirational days may not yet be over. So that all those who so kindly believed in and assisted one 'cranky ageing inventor' and have since passed along the way, may not have done so totally in vain.

Not a car in sight. A somewhat wintry view of Fort Victoria belies its busy holiday season look with many cars and a crowded upper viewing deck. To the right of the signpost can be seen the Planetarium and Aquarium, the Restaurant and the Toilets are out of the picture to the extreme left where we had originally sited them. The Vistarama type museum was to have run through the length of the remaining units.

Design Office display model of the Black Knight ballistic rocket kindly obtained for the author by Raymond Wheeler, Systems Support Director of Westland Aerospace Ltd.

1952. Unwitting participation in a synchronistic social tableau.

The first charade

I have always felt that somehow, somewhere, within the following bizarre accounts there is a message. Whether or not it can be interpreted generally, personally or not at all, is perhaps not for me to speculate.

As is well accepted, working in the environs of a large industrial complex for several years inevitably fosters an exchange of background experience between departmental colleagues. In my case, after I had been working at D.Napiers for some time, several heard of my earlier association with David Kossof, who by then had become a celebrity as a film and TV actor. Again inevitably this must have filtered through to my boss, chief of the Technical Publications division.

One day I had returned from the Projects office to be greeted by several of my pals with the news that the boss was interviewing a VIP, and this person was none other than David Kossof. Sure enough I recognized those familiar authoritative tones from the 'glass cage' confirming indeed this was so. There was David in animated conversation with the boss and several attendant departmental section leaders.

Now it is a fact of life there are some people who live a Walter Mitty existence in which they either falsely claim very close connections with the famous, or they suspect that other people are similarly afflicted. As will be shown later, regretfully I have to say our boss was one of the latter. So that morning I had a distinct feeling I could be a candidate for one of his juvenile call my bluff charades and with due regard to both David and myself I opted for an immediate about turn to vacate the office. Perhaps he anticipated my intention, but the next thing I knew, there framed in his office doorway stood the boss, loudly calling for me to join him, and still with intentional full voice ushered me into the office and introduced me to David saying "Ah Mr.Kossof, I believe you know our Mr.Cramp". Everyone in that office stared expectantly, the outside chatter of voices and typewriters became momentarily still. If the boss had intended to embarrass me, he succeeded. I glanced over to David who only moments ago had been pronouncing in that (to me familiar) authoritative public voice and without a glimmer of recognition he mischievously let me sweat for what seemed to be an eternity before jumping to his feet, proffering me his hand and in his ordinary cockney voice said "Know im, why we were schooled in the same bloody stables, how are you Len?"

Later David and I were able to confer and I said I hoped this little charade hadn't inconvenienced him, but he assured me this wasn't so, for he had already 'sussed that little bunch out' for after all hadn't he been through the same 'school' himself?

**A touch of royal
synchronicity**

In 1953 I had been working at D. Napier for over five years, it would be another three years before my adventure with the cold turbine as discussed earlier. At that time I had been involved with engines of all descriptions, from the well-known Deltic compound IC engine, to turboprops and turbojets. The applications were as varied as they were many, involving locomotives, ships and aircraft (both rotary and fixed wing). Among a resident team of some ten illustrators, engineers and technical authors, two of us were on loan to the Projects office for specialized work. As previously stated, much of this work was urgent, now and again requiring `back engineering' from us.

When I first joined the company one other illustrator had usually been called upon to do this work, but due to promotion this work came to me. Technical illustrating involves perspective line drawings of a multitude of components, which can vary from a few bits of hardware like nuts and bolts, to highly intricate assembled structures. But for most illustrators this seldom involves completely sectioned engines and even less the very large size drawings typically measuring between three and four feet wide. To go one stage further and produce these to an orthographic scale, contoured and sometimes coloured requires very special skills and is often colloquially designated `the cream' by envious workmates.

The technical publications department at Napiers occupied a whole division of the top floor of a very large block of offices. At that time the total work staff consisted of about ten illustrators, three airbrush artists, several typists, seven technical authors, two section leaders, of which I was one, assistant and chief. As described in Chapter 2, usually the drawing boards were roughly divided with some situated along the outer wall and the remainder forming a central aisle. Usually the size of the black and white drawings to be found on any one illustrator's board would measure approximately fourteen inches square, the rest of the space like as not taken up by odd `dyeline' machine drawings. Such was the situation that memorable week in 1953 at which time we were expecting a visit from a VIP.

The day of the visit duly arrived and found me delayed at the Projects office, so I had no idea of what was afoot. On my way back to `Tech Pubs' I had been advised to hurry as our visitor had already arrived. As it turned out I would be grateful later for that moment of grace and I shall never forget the extraordinary sight which confronted me as I rounded the corner into our department.

Followed by his attentive entourage our Very Important Visitor was already half way along the office, pausing interestedly now and again at the various drawing boards, while my illustrator colleagues - seemingly a little embarrassed I thought - stood aside. As he proceeded my eye caught a glimpse of first one and then another familiar rendering and for an agonizing moment I became transfixed in shocked amazement. For on drawing board after drawing board, there in all their oversized glory were examples of *my* work! I could hardly believe my eyes, turned and promptly left the office,

Apparently our well intentioned boss had considered our visitor would find this type of presentation more `impressive', when in fact this work represented a mere fraction of the normal day to day program! My colleagues took this affront good naturedly enough, but I would rather have been spared the embarrassment. Moreover the ensuing tension didn't do much for departmental relations. True this kind of knee jerk behaviour is rather disturbing, but to me, no less intriguing is the fact, that within the synchronistic tenor of this book this wouldn't be the last time (by only a few months as indicated in Chapter 3) that HRH Prince Phillip was to unknowingly cast an appraising glance at some of my work.

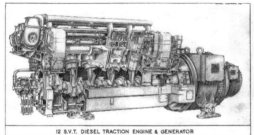

12 S.V.T. DIESEL TRACTION ENGINE & GENERATOR

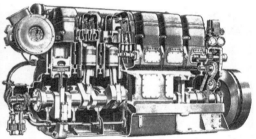

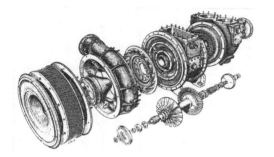

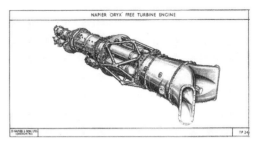

NAPIER 'ORYX' FREE TURBINE ENGINE

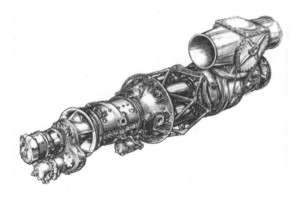

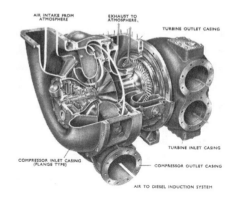

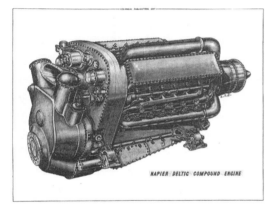

NAPIER DELTIC COMPOUND ENGINE

Typically measuring some three to four feet wide, this sampling of nine drawings was rather chauvinistically distributed around the 'Tech Pubs' office at Napiers on the morning of the visit by a VIP in 1953. But strangely it wouldn't be the last time that HRH Prince Philip would cast a measured glance at some of the author's work.

1940s - present. This work stems from the author's original Ascentrodyne `space' drive, (an early attempt to derive linear thrust from assymetrical angular momentum forces) which eventually has led to some surprising developments.

ATF 55 years.

A culmination of over four years work with the help of the author's friend, John Campbell. It was this work which was referred to in the acknowledgement page of the author's book `Space, Gravity & the Flying Saucer' in 1954. It is interesting to note that this research also included rotating gyroscopic systems which have become popular today.

The enclosed photographs show one of the natural developments of this long inquiry into more mechanistic `space' drives. Results were much the same as those which accompany rotary gyroscopes i.e. an exhilarating dramatic large pulse followed by a series of negating mini-pulses.

Logical considerations of dynamics deem all such devices incapable of producing definitive monodirectional thrust. But it will be shown this may not be the end of the enquiry.

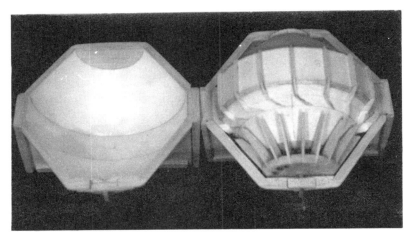

From the early 1940s. A mock up of the author's Ascentrodyne `space drive'

At this point I am acutely aware of the fact that there is one large piece of this jigsaw missing. But it must be remembered it is precisely due to the size of this piece, that it had to be carried on its own. To have published that material in this present volume first, would have been tantamount to putting the cart before the horse. The horse is now happily stabled in the Cosmic Matrix. However we can at least touch on the pedigree of the beast,

By now the reader will be aware of the fact that for most of my adult life I have been actively devoted to the idea that gravity is but another facet of nature waiting to be fully understood and finally overcome. But such a view is so common nowadays and taken for granted, that it might be difficult indeed for some to imagine what it was like for me as a youngster to harbour such thoughts when even rocket propulsion wasn't publicly understood. Today, due in part to films and space adventure stories, the principle of action and reaction has been so commonly discussed that it has almost become natural language to many youngsters. Therefore I have no difficulty in imagining that in time to come young people of the future will be talking the language of anti gravity in a similar vein.

True, that to have an awareness of the possible nature of gravity is an advantage, but the search for the key to unlock it is, to put it mildly, a formidable enterprise. So much so that almost in exasperation one is tempted to return again and again to purely mechanistic solutions. Over the years there have been many attempts to do just that. In my own case John Campbell and I worked over four years investigating various dynamic systems and particle perambulations. Among the 'space drive' adventures were precessing gyroscopes, quantum jerks and centrifugal force, predictably all doomed to failure. The author's enclosed Ascentrodyne was the culmination of one of them.

The space drive

The descriptive term 'space-drive' doesn't necessarily imply space in the planetary sense. It is used by engineers to describe a physical process of manipulating masses to produce a *spacial displacement* , or movement from one point to another by a mechanism. As in the case of a precessing gyroscope is a good example.

Long ago when I was acutely aware of my limitations in the world of particle physics, like so many others, I could only give rein to my dreams about anti-gravity by excursions in pure dynamics as opposed to chemical rockets. Indeed several of my early arguments in this vein were used in Piece for a Jigsaw and they still hold true today. However, always - and no doubt fortunately - the basic laws set down by Newton hold true. So should one be tempted to stray into such dynamical meanderings with experimental hardware, it can assuredly be taken for granted that it will be doomed to failure. In other words it is a time and expensive means of proving the natural system.

Sometimes I have offered the analogy, that if one imagines an otherwise empty physical globe suspended in space containing say ball bearings representing molecules, it follows that no matter how much, or what kind of gyrations they inherit, always the reaction forces are eventually cancelled out. True the globe will be displaced in space, but it will always ultimately return to the starting point..

Precessing gyroscopes follow the same pattern and any apparent forces are ultimately cancelled out. True, sometimes the movements involved appear to be so complex that one can be tempted to try it and see, just as I did, many times. The Ascentrodyne shown here was the culmination of such efforts.

However despite the disappointments, such inquiry can sometimes pay very useful and unsuspecting dividends. As in fact was the case for me several years ago. Unfortunately this work is at present classified and cannot be discussed, suffice to add that any of the foregoing vehicles, from the CM 100 on - which were primarily designed to operate with various multiple engines/fan systems - could now incorporate an entirely novel lift/thrust system, tentatively titled SDU 15.

As with the electromagnetic and gravitic versions described in the Cosmic Matrix, such units will be electrically powered, practically silent, use a fraction of the energy normally associated with internal combustion engines and develop no pollution. But its sphere of operation will be global rather than interplanetary. The age of the 'floater' has all but arrived. It is fervently hoped that this technology, this fish, will prove to be more digestible for a race of people who appear to suffer from over sensitive stomachs!

In other words, although such technology is now becoming pressingly necessary, even this would have been economically unacceptable only two decades ago. As in fact it seems are my conclusions on gravity and energy economically and politically unacceptable today.

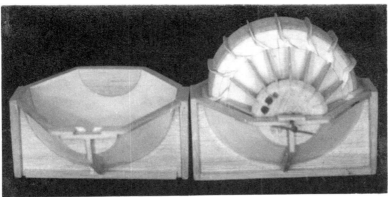

Two further views of the early Ascentrodyne, thrust was measured through the longitudinal axis. Although unsuccessful, over the years this mockup has acted as a constant stimulus to further space drive schemes the latest derivative of which shows significant promise.

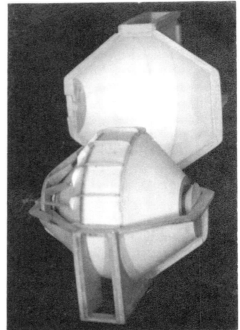

The frontispiece to this book is emphatically not a fantasy situation, but based on this present work, a genuine forecast of what assuredly can come. It depicts a version of the author's CMAL series craft in silent hover. It employs no conventional lift/propulsion matching engine systems, and employs comparatively small electric motors, together with an automatic solid state three axis control package. If allowed, this can represent the transport technology of the immediate future.

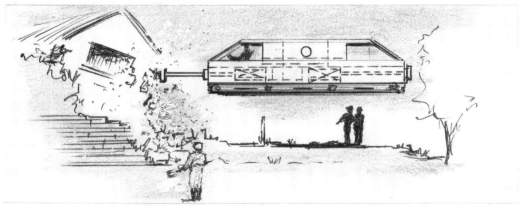

Two of many predictable scenarios in which vehicles operated by the author's SD 15 units could be used, whence there would be an absence of dust or 'whiteouts' associated with helicopters.

From miniature gas turbines to hovercraft,hoverplanes, UFOs,anti-gravity and star ships in the sky

In general summary to this autobiographical report, it is my hope that the foregoing information may prove to be of general interest, if not in some measure useful to those who would pursue similar paths. Also in offering this somewhat convoluted preamble with some of its thought provoking mysterious connotations, I ask forbearance, in so far that I am just as puzzled by some of the bizarre aspects as anyone else would be, though of course I have my suspicions. However I hope the inclusion of some of the previously published material here will offer a measure of authenticity, as it can be substantiated as being perfectly true.

Should there be any remaining doubt about the reality of synchronicity and/or fore-sight there are other worthwhile specific sources of information. In this book I have of-fered only a few cases known to me, but there are many others such as the following. In *July* 1975, one Erskine Lawrence Ebbin was fatally knocked off his moped by a taxi in Hamilton, Bermuda. It was the same taxi with the same driver, carrying the same passen-ger, which had killed his brother Neville, in *July* the previous year. Both brothers were seventeen when they died, riding the same moped in the same street.

In 1956 I went for my first glider trip. I was feeling very nervous, and we had *three abortive take offs* due to the tow hook malfunctioning. Years later I went on my first flight in a Trident aircraft, *we had three abortive take offs* due to the rear engine failing to start, both pilots involved claimed that this had never before happened to them.

As I have said, in my case the events in this book are *in themselves* of relatively little significance, while the *number* of them spread over a *lifetime* renders the numerical chance odds `explanation' rather tiresomely thin.

Test rigs and models adorn a corner of the old workshop.

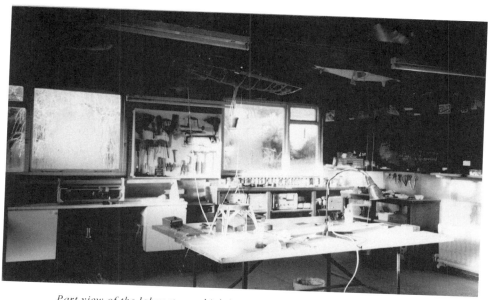

Part view of the laboratory which has seen the creation, but sadly not the fruition, of many extraordinary things.

NOW, THE HOVERPLANE!

SOUTHERN EVENING ECHO, Thursday, February 24, 1983   Page 5

New development from home of the hovercraft

By Maurice Leppard

THE ISLE OF WIGHT — home of the hovercraft — is poised to give birth to another leap forward in travel . . . with the hoverplane.

Inventor Leonard Cramp is confident he has perfected the design for the new craft and is ready to switch from working models to a full-sized prototype.

For Mr. Cramp (63), of Cranmore, near Yarmouth, the breakthrough comes after 20 years of research in the workshop at his secluded home.

He believes from experiments with models that his hoverplane, which in one role looks like a flying saucer, can fill the gap between the hovercraft and helicopter.

"The hoverplane is an aeronautical hybrid," says Mr. Cramp. "It combines the heavy load carrying advantages of the air cushion vehicle with vertical take-off and landing capabilities of the helicopter."

In its freight carrying role he envisages huge hoverplanes transporting cargo pallettes over rough terrain or water, able to fly over buildings, forests or other obstacles.

One major British company is already seriously considering this

MR. CRAMP pictured with some of his working models.

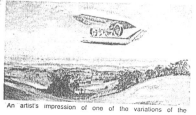

An artist's impression of one of the variations of the

A Southern Evening Echo article in February 1983 sums it all up.

Appendices

Appendix 1

The Inventor and the system

For those who would wish it, the following is based on a hard core of facts as I have found them since the late 1940s and for convenient assessment a little of this material was duplicated in the Cosmic Matrix. Today there are signs that in the UK there is hope of a better deal for would-be inventors. Even lay people are becoming aware of the fact that all too often highly trained professionals are usually so tied to *directed* research in industry, that although they have the means, they seldom have the time to conduct *original* research. Therefore there is no mystery in the fact that many significant technological developments are often initiated by a few freelance individuals such as the author.

The inventor

Usually, most inventors are not of the ilk who clamour after acclaim and `banner waving', the exhilaration of success is more akin to a stroller on a beach finding a most beautiful stone, and having experienced the first reaction of delight, he or she then wishes to share the find with somebody else. There is no predominant sense of `look how good I am' about that situation, and most inventors behave in a similar way. For an inventor to behave with overbearing pride and self glorification is most uncharacteristic of the species, its a failing which merely betrays a juvenile status. It would be tantamount to the above beach stroller having found the pebble, covetously showing it off as if the inventor was the something special.

In fact all too often with most true inventors the reaction of achievement is the reverse. They are usually the type who suffer more than a little embarrassment with lavish public acclaim. Also they are so involved with their wild schemes and ideas that they can be very impatient with trivia, neither do they suffer scoffers lightly.

But like it or not, we do live in a world where there are many scoffers which is not entirely a trait of the layman. It is from this quarter that the inventive or artistic soul can receive his most stinging barbs. These are the type who will instinctively frown on an agreeable painting or exciting invention and even treat the artist/inventor as if they were something unclean or had the plague. I do not exaggerate, I have literally seen such a person - who held a very senior position in a large company - contortingly turn his back on an obviously nice exhibit in a situation when it wasn't a physically comfortable thing for him to do. In this instance there were others present who witnessed this extraordinary behaviour from an otherwise intelligent man. No doubt psychiatrists have a name for it and so have I. But the disturbing thing is that many of his type find their way into the higher echelons of our society.

It is interesting - albeit bewildering - to note that this type of reaction from such individuals varies in degree in direct proportion to the import of the `exhibit'. I wish to avoid any responsibility of judgement in this respect and am content to leave to the reader's imagination on how the material in the Cosmic Matrix and this present book will be received by such elements.

The inventors lot

anyone who has ever conducted unsponsored private research knows only too well that there can be many pitfalls. For instance as has been shown, with very limited means my family and I have contributed valuable time to the enclosed and other related research. Sooner or later it has to stop. To continue the work at a meaningful level requires unattainable funding; significant for the individual, petty cash by even a modest firm's standards, but miniscule when assessed in terms of national or even state university level. People ask `why not apply for a university grant?' Sober reflection will reveal that this can be an irresponsible thing to do. The Department of Industry, British Aerospace or Ministry of Defence perhaps?. From experience the author knows only too well what that would mean. For those unacquainted, it would be very much a case of the `don't call us' syndrome. That is of course unless finished, fully tested and confirmed hardware is put on their table, and *then* have little doubt, the presenter would probably experience great difficulty in getting himself and `it' out of the door1

One of the great problems for many people who have ideas of their own is the `we own you' clause found in mandatory agreements between employees and most larger companies.

In essence, the argument that employees might produce a new idea which may have been directly or indirectly influenced by the nature of their work in the company's related end products, is obviously a fair one and most people I have met agree as much. However, as always in order to show their petty worth, there are those bigots who are inclined to spoil this world by carrying matters to stupid extremes. For example, as has been shown, many years ago when I conducted my private work on small jet propulsion gas turbine engines, due to scale effect problems this required special solutions. Then *years later* when I joined D.Napier who had their own problems with high temperatures to solve, a solution I had those years before was rekindled and I offered it to them. Despite my assumption that this idea was my property and in no way could have been the result of this company's products, they were quick to `adopt' it and patent it. Sure, they did give me *credit* as the inventor, but *officially* had they developed the design they would not have been obliged to give me a penny piece! I would have liked to believe that this would have been reflected in my salary, but I doubt it. As stated, subsequently that company became insolvent and that too was many years ago, but as we have recently seen a strangely familiar design is being contemplated for a new space craft of the 1990s. Lest there be any suspicion that the present author is biased or carries an oversized chip, may I hasten to offer the following example of this bureaucratic insensitivity.

As shown in the last chapter in a furtive attempt to increase my already stretched income, my sons and I became involved with Lesneys to develop some new ideas. Many of these were patented and are now well known the world over, for which we have never received any royalties! No less to the point however concerns this instance which occurred while I was employed at the British Hovercraft Corporation on the Isle of Wight in 1965.

A good friend there - knowing of my association with toy manufacturers - came to seek my advice concerning his very elegant solution for manually steering toy cars. It should be emphasized that the nature of this person's work had nothing to do with aircraft *design* of any kind and in no way could his idea have been influenced by this company's products. Nevertheless I agreed with him that as a formality he should go 'through the system' before taking out a patent. The patent officer not only took possession of the idea, where it was *shelved for months,* but further insisted it *might* just have a suitable application for aircraft! In fairness I should add that in this instance the patent officer was not a professional in that capacity, so I am pleased to relate that on my recommendation my friend was able to retrieve his model and specification and ultimately patent and profitably sell his idea, which was manufactured by Mettoy UK. This toy was sold all over the world, but here again no thanks was due to the 'more than my job's worth' types in the establishment!

I have to confess, this problem was an additional reason which prompted me to leave a good position in one of the largest sections of the UK Aerospace industry and take up another position with a smaller aircraft manufacturer who had more things to do actually building aircraft, than to deny an individual the right to indulge in new ideas.

The establishment

It would be nice to believe that the day will dawn when the establishment - particularly in the UK - will openly recognize the fact that a significant proportion of the major steps forward in technology have been made by *non* establishment individuals, some of whom were regarded by them as having little more than sophisticated amateur status. For years the establishment has bathed in the reflected glory of these people, when we 'in the trade' know only too well their work had been initially brushed aside. I list but a few. Herman Oberth and his work on rocketry, R.Hook practically unknown *original* inventor of the modern hydrofoil, John Logie Baird inventor of the first disc television system, Orville and Wilbur Wright for the first successful man carrying aeroplane, Igor Sikorski forerunner of the first successful helicopter, Sir Frank Whittle inventor of the first successful jet propulsion gas turbine engine, Sir Christopher Cockerell inventor of the peripheral jet hovercraft, etc., etc.

The other side of the coin

But if we are to be allowed the above chaffing of the Establishment, perhaps it is only fair to redress the balance somewhat by adding, that with the utmost respect, the author finds it very difficult to resist a little - perhaps cynical - smile when reading of the *difficulties* that some well known and prominent inventors claim to have met. And from a comparatively lowly position it is nonetheless - at times - irksome to reflect on issues such as the following. When speaking of his early efforts to get recognition and backing for his work on wave energy, the late Sir Christopher Cockerell - having been disappointed by a no response attitude from Mr.Goodwin at the Department of Energy - carried on, using his own resources, before finally the British Hovercraft Corporation 'out of the kindness of their hearts' agreed to make and test some models for him. Such models built by the establishment would normally have cost thousands of pounds, but BHC could hardly have done otherwise seeing how this brilliant man had practically rescued them! Even so, Sir Christopher expressed - perhaps rightly so - indignation over the fact that Stephen Salter inventor of the Duck system for wave energy (who was an apprentice when I was at Saunders Roe) had the facilities of State University and government backing *before* the official grant was made... While Mr.Goodwin said that "Salter is inclined to make fools of people who had greater facilities than he enjoyed".

Thus we have a very rare situation in which one inventor *was* getting privileged help at the tax payers expense - both funding, assistance *and* university facilities - and another similarly from a large aviation corporation and *both* feeling more than a little hard done by! The author humbly suggests that with the passage of time there is a risk of losing one's sense of propriety and that it should be remembered, that the above style of *considerable* help so few people enjoy, is denied to most who are *totally* without meaningful support or facilities of any kind. At the time of writing some of us 'out here' still know what it's like to make our scalpels from worn out hacksaw blades, *design, construct and test our own models,* because we can't afford to get others to do the job for us. We still know what it is to sweep out the workshop, share strip lights, or even go without and work with one 60w lamp. Climb on to the workshop roof many times to stop the annual bins full of rainwater ruining our experiments. Repair the car in order to pick up visitors. Not least, feeling guilty about the amount of typing, book keeping and general running about of a much overworked, uncomplaining helpmate and wife. And all this not for just a couple of years, but over a lifetime! Need I go on, or perhaps I have made the point. There are many people like the author, but it looks as though there are a few others who either have short memories, or perhaps when it comes to *really* roughing original research, they have been very fortunate.

It is not only relevant, but vitally important to the main intent of my books that people are made aware of the extraordinary 'inertia' towards expanding technological development which exists at certain levels in the world. Indeed it not only exists, but hindsight reflection is inclined to make one marvel that we *ever* got the technological status we enjoy today. The acceptance and eventual use of the technology I have revealed in the Cosmic Matrix will depend solely on this veil of cynicism and selfishness among some of our species being penetrated. There can be no more effective way of doing this than by making people aware of the situation through the experience of others and in this context it is hoped that the reader has a sense of humour.

As shown, since the early days in a private capacity, I have 'brushed shoulders' with the Establishment, Industrialists, Politicians, Innovators, yes including con men, the Toy industry and publishing. This somewhat checkered career has sometimes been envied by some of my friends, who have reckoned their lives to have been mundane by comparison. I have hastened to point out, that's OK if you are prepared to take the blows, ridicule and near poverty this little lot can bring. I did agree with them however when they said it could fill another book!

The band wagon jumpers

Also as previously stated during the years I worked on various projects, I used my own limited income to foot the bills. Not surprisingly therefore the time had to come when on several occasions I was glad to respond to offers of sponsorship from individuals previously unknown to me. Now although it is easy to be wise in hindsight and that caution should be the keynote, it is very difficult to refuse when *survival* is looking at you across a table with an open cheque book in its hand.

Usually such people are very pleasant and quite sincere, but they are also very hard headed business types out for a quick profit. That is fair enough, except that the associated conditions are not. At the end of the day what it really amounts to is, first, most of them are inadequately financially viable, and therefore the 'funding' means a pittance of an income for an average weekly involvement of 120 hours (the inventor is his own slave driver). Secondly, no matter how many years, or how much of the inventor's own money may have been originally invested, the procedure is to take the idea to a viable status to put on the market, from which income sponsors are apt to insist on *first recouping their funding.* Thus they are involved at the rate of perhaps 65 - 35% shares (if you are lucky) thereby having control of any ensuing company, usually on the ploy that most inventors are not good business people, while *they* are. Unfortunately I and others have found this to be the other way round. In the final analysis this means they have used the inventor's ideas, his time and his money to get into a project over which they will have dominance scot free! It is astonishing that they seem to be unable to appreciate this obvious truth. Extraordinarily enough, when the next in line entrepreneur approaches the inventor, they show the most genuine heartfelt alarm that he should have been so treated by their predecessor, then proceed to do exactly the same thing quite oblivious to the same fault in themselves!

Not surprisingly therefore such ill founded enterprises usually come to nothing, sometimes leaving the inventor in debt and bewildered, a little wiser and certainly *always very disappointed.* I am no longer surprised - although sometimes taken off guard - by such behaviour, but I am often alarmed by the thought as to how much their kind in all quarters of innovation, may have been directly responsible for delaying technological development rates on a world wide scale.

Knighthoods & all that

Unfortunately there is another facet to this tarnished coin, for on occasion when officials - no doubt frantically responding to popular media fervour of the time - *does* decide to acknowledge innovation, it promptly betrays its utter incompetence concerning internationally related 'state of the art' matters and quite oblivious to overseas sensitivity, wallows in the backwash of similarly misinformed media reports.   Thus a potential Knighthood situation can be born, as we saw in Chapter 7. So, with all due respect, for one converted coffee tin experiment and one small model, built by a handy neighbour, plus a few covering sums, Sir Christopher didn't do so bad with a knighthood plus a sum of money - which even by today's standards would be very welcome news indeed to most of us - even if he did have to wait some *twelve years* for it!

In a like manner Sir Frank Whittle has gone down in history as being the *inventor* of the jet propulsion gas turbine engine per se for which he received an ex-gratia payment - not much - and again a knighthood. Yet as some of the older engineers of that time are fully aware, there were other gas turbine designers including the Germans, who were *flying* gas turbine jet propelled aircraft in World War 11 *before* ours first took off. Therefore we can say with a degree of certainty that the device was *simultaneously invented* in Britain, Germany and Sweden, but *patented* in the UK.

As with so many other inventions of both ordinary and 'breakthrough' status, it can safely be anticipated, whatever the subject, that such work is *coincidentally* and quite independently going on throughout the world. So much is this evident in historical records, that I have philosophically stated it is as though the divine authority 'allows' the *next stage* inspiration to filter through when it is deemed mankind is ready for it. Therefore quite suddenly thousands of people simultaneously get a 'new' idea for some purpose. A few hundred may be motivated to actually do something about it, but for various reasons drop out along the way.   The few remaining dozens, usually poor, succumb to the 'private funding grabbers' and/or the establishment, while the remainder - one or two in different countries - through good fortune (such as being in the right circumstances at the right time) and or staying power, make it and a potemtial patent is born.

I pen these words in the full knowledge that somewhere out there, there is another 'me' who over the years has had similar experience in his efforts towards a new age energy/

propulsion system, only this time for some obvious reasons and others not so, it may *never* be patented.

It must be reemphasized that absolutely no discredit is intended towards Sir Frank Whittle, Sir Christopher Cockerell or anyone else whose work has furthered technological development, but I welcome this opportunity to speak up, not only for myself, but for the host of other engineers whose dreams, money and work - perhaps over a much longer period of time - go completely unheralded. To say that this is because their work is not published is not true, as I hope I have illustrated, much of it is. But beyond any doubt, at the end of the day it is the chap who has the means and gets to the patent office first, no matter how brief or recent his work, who gets recognition and all that goes with it. It's a pretty harsh system and about time we took a lesson from our American cousins and set up an institution which will help inventors freely. At the time of going to press I am pleased to say it looks as though this is now beginning to happen in Britain.

The author like many others, would like to be assured that he *does* live in a society that respects and acknowledges innovation and puts its money where its mouth is to confirm it, rather than a system which - if for instance my work is published - will acknowledge this attempt to raise further research funding, merely by imposing heavy restrictive tax burdens on any royalties we might earn, *over the immediate future,* when I have devoted *much of my life* at considerable expense. This is true, novelists who may have spent a lifetime and expense to reach viable status, on publishing their *first* book, used to be taxed over the *first* year. It has now been graciously extended over three years!

I repeat, I have no wish to discredit, contest, or malign eminent inventors or scientists in any way, but I do wish to reveal a system which is self evidently set up and manipulated by people of little more than amateur status, who regard themselves fit to make important decisions about technological developments of the above kind, and risk making a laughing stock of our country in the process.

It may be difficult for some lay people to visualize this type of official incompetence in society, but it is not only prevalent, it comes in all guises. Not surprisingly most of those who participate in it, more often than not, are quite blameless in the sense that they are totally unaware of the *role* they appear `programmed' to take.

For instance, should a minister at some level or another be detailed to investigate the validity of this author's work, probably not being a trained scientist, he would hardly be in a position to do so. Therefore he would have to seek the advice of some `authority' in the establishment, thus placing the future of the work in jeopardy by relying on the opinion of a traditionally entrenched - and like as not biased - scientist of the type I introduced earlier.

Naturally, after having devoted years and expense to establish credibility in such work, this would be a disrespectful affront and I personally would resist such intervention vigorously. They would have to judge my work on merit shown or disregard it, as they often do. Should there be any doubt about this I would remind the reader that I worked at the establishment, where they treated Cockerell as if he were a nuisance with a stupid idea, rather than someone who had achieved something that *they* should have. Then went on to adopt his work and make a handsome profit from it. This would no doubt be conveniently denied by some, but I have met Cockerell and know it to be true.

Any reasonably intelligent person today intuitively feels that the technological developments predicted in the Cosmic Matrix could happen. Aircraft, cars, spacecraft, in fact all our transport - if allowed - will be different in the future to what they are today. Most people feel at the back of their minds that our energy systems will be different, the nature of which they are uncertain. Therefore in this respect I will be telling them nothing new, except that it can probably happen in the near future. The big question that this message poses is; are we ready, or for that matter even fit for it? It is in part due to that unanswered question that I have devoted my efforts to produce the aforementioned `halfway house' alternative of the SDU system.

High profile trend setters

By way of summary to this Appendix, past experience prompts me to forcibly reiterate the difficult task of establishing credibility when one represents a very small voice among the crowd. This truth is no less so when the leaders of the pack are many and have the loudest voices and it is in this context that readers should be forewarned of `official' denigration to new ideas. As heavily portrayed, during my life I have encountered many instances of how these `lions' of society make fools of the rest of us. However with the passage of time even the most humble of souls benefit somewhat from such experiences, so that the false ones are more easily recognized. A couple of more obvious examples I know of will not be amiss:

Case 1. Mr. L. a stockily built man of medium height, a professional industrial artist. He had short fingered powerful hands, more reminiscent of a builder's labourer than an artist. Wore a cloth cap, had a quiet unobtrusive manner and inclined to be a little shy,if not introverted. But he could swim like a fish and paint like a master! Indeed after having worked close to him for a number of years, I only discovered privately that he had been approached to exhibit at the Royal Academy. Even then he was rather embarrassed to admit as much.

Case 2. Mr. B. a charming enough person, fairly tall and of imposing appearance, dearly *wanted* to be an artist. He dressed the part a little extravagantly, complete with cravat, white smock and vandyke beard. An extrovert he would often expound loudly and authoritatively on matters of art, particularly impressionism. The illustrative point being, of the two of them, Mr. and Mrs. Lay public were more likely to regard the latter as being the *expert.*

My second example, also from the arts, is a recently revealed statistic which I suppose many of us felt in our bones. The fact that of the hundreds of known orchestral conductors of highly paid rank, only a mere handful are of meritorious competence, many of the remainder - being noticeably more artistically flamboyant - are often of little more than jumped up amateur status. Extraordinarily enough, most of the musicians who actually do the work are often by comparison lowly paid. Indeed it is not unusual for conductors to receive more than their entire orchestra. Again, these are the establishment *experts* which dominate the scene and the voice of *authority* mostly heard by members of the public, and the situation often dominating the scientific establishment is exactly the same. In a word we should automatically be on our guard when loud voices of authority proclaim and that `by their fruits we shall know them'. Unfortunately for us all, any of us wishing to gain informative advice on any topic have little other recourse than to consult the experts. The moral has to be, do not rely too much on the loudest of them!

For my part I have never considered myself in that particular role, rather I would like to be thought of as none other than that little kid with a love for all things beautiful and those that fly and who always had an insatiable desire to know what makes things tick. It will be left to the readers of my books to judge if in the fullness of time I ever made it or not.

Appendix 2
Chronology of surface effect vehicles

The following is an extract of information given to the author by the Patents Officer of the British Hovercraft Corporation many years after the launch of the SRN 1 hovercraft 1959.

1716   Emmanuel Swedenborg, Swedish designer and philosopher, proposed a vehicle rather like an upturned dinghy, with a cockpit in the middle. Apertures on each side allowed the operator to raise and lower a pair of oar-like scoops, which on the downward stroke *forced air beneath the hull, thereby raising it above the surface.* Project was abandoned when Swedenborg realized the machine would require far more energy than that which could be supplied by a single person.

1860s   General proposals for admiralty skimming vessels.

1874   Lord Thorneycroft began experiments with `air-lubricated' hulls and by 1875 correspondence on possible applications was being exchanged between admiralty contractors in the United Kingdom and the Netherlands.

1897   During this twenty three year lapse literally hundreds of British, American and Swedish inventors were hard at work on these lines, but none appear to have produced a successful working design, until an American - a Mr. Cuthbertson - filed a patent in which he forecast the configuration of today's `sidewall' air-cushion vehicle (ACV) with surprising accuracy.

1909   Hans Dinesou, a Swedish engineer, completed the first detailed design of a sidewall ACV with flexible fore and aft rubber `cushion' seals.

1916   Dagobert Muller von Thomamhul, an Austrian engineer, designed and built a sidewall ACV torpedo - `boat' for the Austrian navy. It was said to have reached speeds of over 40 knots and as far as it is known was the world's first successful ACV.

1930   T.J.Kaario of Finland - who is still conducting ACV experiments - attained a speed of 12 knots over ice with his first ram-wing single seater.

1939   With the above exception during this twenty three years, again there were many unsuccessful attempts, due not so much to lack of ideas, but lack of financial backing, until A.V.Alcock, an Australian, built a series of successful working models. It was he who coined the term `floating traction' for this kind of transport.

1957   The next twenty years saw a continual development in `open plenum' type ACVs until Christopher Cockerell's achievements were reported by the world's press. Like most of the earlier pioneers he also began by exploring the use of air-lubrication for boats, beginning with a punt, then a 20 knot ex-naval launch as a test bed. The limitations of this soon became apparent and before long he was working on far more ambitious ideas. Unlike some of the earlier plenum ACVs, Cockerell's system employed the now famous annular jet or nozzle which formed the air `curtain' which served to trap the higher pressure air cushion. It was this ingenious idea that initially enabled his craft to operate at greater heights. From then until the present, the hovercraft has been further exploited to a more significant extent in America, Canada, Great Britain, France, Holland, Russia and China.

Author's professional background

DE HAVILLAND AIRCRAFT CORP., HATFIELD.
Mosquito, Hornet, Comet, Vampire aircraft.

D. NAPIER & SONS, LONDON.
Forward Projects Dept., Deltic Compound Engine, Eland & Oryx gas turbine engines etc.

SAUNDERS ROE LTD., ISLE OF WIGHT.
Mixed unit supersonic aircraft SR53, SR177, Scout helicopter, Black Knight/Blue Streak rocket program etc.

BRITISH HOVERCRAFT CORP., ISLE OF WIGHT.
The first SRN1 Hovercraft through SRN2 - SRN6 Forward Projects Dept.

N.D.NORMAN AIRCRAFT, ISLE OF WIGHT.
Firecracker aircraft design team with Desmond Norman 1977.

AIRBILT LTD.
Chief designer research dept.

LESNEY MATCHBOX.
Projects design consultant & model manufacture.

METTOY & TIMPO TOYS.
Projects design consultant & model manufacture.

Bibliography

Adamski, G. *Inside the Space Ships.* Arco publishers. 1956

Baker, A. *The Encyclopaedia of Alien Encounters.* Virgin. 1999

Catoe, Lynn E. *UFOs and Related Subjects.* Published by Library of Congress Science & Technology Division and the Air Force Office of Scientific Research, USA. 1969

Cramp, L.G. *Space, Gravity & the Flying Saucer.* Werner Laurie. 1954

        Piece for a Jigsaw, Somerton Pub. 1966.   Adventures Unlimited 1996

        The Cosmic Matrix, Adventures Unlimited. 1999

Dunne, J.W. *An Experiment with Time.* Faber. 1927

Elsley & Devereux. *Hovercraft Design & Construction.* David & Charles. 1967

Forman, J. *The Mask of Time.* Macdonald & James. 1978

Fuller, J.G. *The Ghost of 20 Megacycles.* Souvenir Press. 1985

Good, T. *Alien Base.* Century. 1998

        & Zinnstag, Lou. *George Adamski: The Untold Story.* Ceti Publications. 1983

Gribbin, J. *Time Warps.* J.M.Dent & Sons Ltd. 1979

Hawking, S. *A Brief History of Time.* Bantam Press. 1996

Hoerner, Dr. S.F. *Fluid Dynamic Lift.* Dept. of Navy USA. 1985

        co-author H.V. Borst.

Hoyle, Sir F. *The Intelligent Universe.* Michael Joseph. 1984

Hynek, Dr.J.A. *The UFO Experience. A Scientific Enquiry.* Ballantine Books. 1974

Kemp, J. *Hovercraft, The Constructors Guide.* The Hoverclub of Gt.Britain Ltd. 1987

Leroy, O. *Levitations.* Burns, Oates. 1928

Leslie, D. & Adamski, G. *Flying Saucers have Landed.* Futura. 1953

Lethbridge, T.C. *The Power of the Pendulum.* Penguin. 1997

Pond, D. *Keely's Secrets, 'Universal Laws'.* The Message Co. 1990

Pope, N. *Open Skies Closed Minds.* Simon & Schuster. 1996

Sagan, C. *The Demon-Haunted World. Science as a Candle in the Dark.* Headline. 1996

Smith, Dr.P.L. *Stability & Control of Hovercraft.* Dept. of Industry,UK. 1980

Streiber, W. *The Secret School.* Simon & Schuster. 1997

Sweetman, B. *Aircraft 2000.* Hamlys Publishing Group Ltd. 1984

Vallee, J. *Confrontations: A Scientist's Search for Alien Contact.* Ballantine Books. 1990

Walters, E & F. *The Gulf Breeze Sightings.* William Morrow & Co. 1990

Index

Sender

Leonard G. Cramp,
Green Acres,
West Close,
Cranmore,
Yarmouth,
Isle of Wight,
PO41 OXT,
UK

COSMIC MATRIX
Piece for a Jig-Saw, Part Two
by Leonard G. Cramp

Leonard G. Cramp, a British aerospace engineer, wrote his first book *Space Gravity and the Flying Saucer* in 1954. Cosmic Matrix is the long-awaited sequel to his 1966 book *UFOs & Anti-Gravity: Piece for a Jig-Saw.* Cramp has had a long history of examining UFO phenomena and has concluded that UFOs use the highest possible aeronautic science to move in the way they do. Cramp examines anti-gravity effects and theorizes that this super-science used by the craft—described in detail in the book—can lift mankind into a new level of technology, transportation and understanding of the universe. The book takes a close look at gravity control, time travel, and the interlocking web of energy between all planets in our solar system with Leonard's unique technical diagrams. A fantastic voyage into the present and future!
364 PAGES. 6X9 PAPERBACK. ILLUSTRATED. BIBLIOGRAPHY. $16.00. CODE: CMX

UFOS AND ANTI-GRAVITY
Piece For A Jig-Saw
by Leonard G. Cramp

Leonard G. Cramp's 1966 classic book on flying saucer propulsion and suppressed technology is a highly technical look at the UFO phenomena by a trained scientist. Cramp first introduces the idea of 'anti-gravity' and introduces us to the various theories of gravitation. He then examines the technology necessary to build a flying saucer and examines in great detail the technical aspects of such a craft. Cramp's book is a wealth of material and diagrams on flying saucers, anti-gravity, suppressed technology, G-fields and UFOs. Chapters include Crossroads of Aerodymanics, Aero-dynamic Saucers, Limitations of Rocketry, Gravitation and the Ether, Gravitational Spaceships, G-Field Lift Effects, The Bi-Field Theory, VTOL and Hovercraft, Analysis of UFO photos, more.
388 PAGES. 6X9 PAPERBACK. ILLUSTRATED. $16.95. CODE: UAG

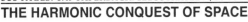

THE HARMONIC CONQUEST OF SPACE
by Captain Bruce Cathie

Chapters include: Mathematics of the World Grid; the Harmonics of Hiroshima and Nagasaki; Harmonic Trans-mission and Receiving; the Link Between Human Brain Waves; the Cavity Resonance between the Earth; the Ionosphere and Gravity; Edgar Cayce—the Harmonics of the Subconscious; Stonehenge; the Harmonics of the Moon; the Pyramids of Mars; Nikola Tesla's Electric Car; the Robert Adams Pulsed Electric Motor Generator; Harmonic Clues to the Unified Field; and more. Also included are tables showing the harmonic relations between the earth's magnetic field, the speed of light, and anti-gravity/gravity acceleration at different points on the earth's surface. New chapters in this edition on the giant stone spheres of Costa Rica, Atomic Tests and Volcanic Activity, and a chapter on Ayers Rock analysed with Stone Mountain, Georgia.
248 PAGES. 6X9. PAPERBACK. ILLUSTRATED. BIBLIOGRAPHY. $16.95. CODE: HCS

THE ENERGY GRID
Harmonic 695, The Pulse of the Universe
by Captain Bruce Cathie.

This is the breakthrough book that explores the incredible potential of the Energy Grid and the Earth's Unified Field all around us. Cathie's first book, *Harmonic 33*, was published in 1968 when he was a commercial pilot in New Zealand. Since then, Captain Bruce Cathie has been the premier investigator into the amazing potential of the infinite energy that surrounds our planet every microsecond. Cathie investigates the Harmonics of Light and how the Energy Grid is created. In this amazing book are chapters on UFO Propulsion, Nikola Tesla, Unified Equations, the Mysterious Aerials, Pythagoras & the Grid, Nuclear Detonation and the Grid, Maps of the Ancients, an Australian Stonehenge examined, more.
255 PAGES. 6X9 TRADEPAPER. ILLUSTRATED. $15.95. CODE: TEG

THE BRIDGE TO INFINITY
Harmonic 371244
by Captain Bruce Cathie

Cathie has popularized the concept that the earth is crisscrossed by an electromagnetic grid system that can be used for anti-gravity, free energy, levitation and more. The book includes a new analysis of the harmonic nature of reality, acoustic levitation, pyramid power, harmonic receiver towers and UFO propulsion. It concludes that today's scien-tists have at their command a fantastic store of knowledge with which to advance the welfare of the human race.
204 PAGES. 6X9 TRADEPAPER. ILLUSTRATED. $14.95. CODE: BTF

MAN-MADE UFOS 1944—1994
Fifty Years of Suppression
by Renato Vesco & David Hatcher Childress

A comprehensive look at the early "flying saucer" technology of Nazi Germany and the genesis of man-made UFOs. This book takes us from the work of captured German scientists to escaped battalions of Germans, secret communi-ties in South America and Antarctica to todays state-of-the-art "Dreamland" flying machines. Heavily illustrated, this astonishing book blows the lid off the "government UFO conspiracy" and explains with technical diagrams the technology involved. Examined in detail are secret underground airfields and factories; German secret weapons; "suction" aircraft; the origin of NASA; gyroscopic stabilizers and engines; the secret Marconi aircraft factory in South America; and more. Introduction by W.A. Harbinson, author of the Dell novels *GENESIS* and *REVELATION*.
318 PAGES. 6X9 PAPERBACK. ILLUSTRATED. INDEX & FOOTNOTES. $18.95. CODE: MMU

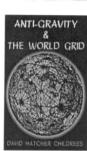

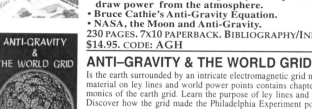

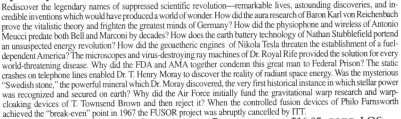
FREE ENERGY SYSTEMS

LOST SCIENCE
by Gerry Vassilatos
Rediscover the legendary names of suppressed scientific revolution—remarkable lives, astounding discoveries, and incredible inventions which would have produced a world of wonder. How did the aura research of Baron Karl von Reichenbach prove the vitalistic theory and frighten the greatest minds of Germany? How did the physiophone and wireless of Antonio Meucci predate both Bell and Marconi by decades? How does the earth battery technology of Nathan Stubblefield portend an unsuspected energy revolution? How did the geoaetheric engines of Nikola Tesla threaten the establishment of a fuel-dependent America? The microscopes and virus-destroying ray machines of Dr. Royal Rife provided the solution for every world-threatening disease. Why did the FDA and AMA together condemn this great man to Federal Prison? The static crashes on telephone lines enabled Dr. T. Henry Moray to discover the reality of radiant space energy. Was the mysterious "Swedish stone," the powerful mineral which Dr. Moray discovered, the very first historical instance in which stellar power was recognized and secured on earth? Why did the Air Force initially fund the gravitational warp research and warp-cloaking devices of T. Townsend Brown and then reject it? When the controlled fusion devices of Philo Farnsworth achieved the "break-even" point in 1967 the FUSOR project was abruptly cancelled by ITT.
304 PAGES. 6X9 PAPERBACK. ILLUSTRATED. BIBLIOGRAPHY. $16.95. CODE: LOS

SECRETS OF COLD WAR TECHNOLOGY
Project HAARP and Beyond
by Gerry Vassilatos
Vassilatos reveals that "Death Ray" technology has been secretly researched and developed since the turn of the century. Included are chapters on such inventors and their devices as H.C. Vion, the developer of auroral energy receivers; Dr. Selim Lemstrom's pre-Tesla experiments; the early beam weapons of Grindell-Mathews, Ulivi, Turpain and others; John Hettenger and his early beam power systems. Learn about Project Argus, Project Teak and Project Orange; EMP experiments in the 60s; why the Air Force directed the construction of a huge Ionospheric "backscatter" telemetry system across the Pacific just after WWII; why Raytheon has collected every patent relevant to HAARP over the past few years; more.
250 PAGES. 6X9 PAPERBACK. ILLUSTRATED. $15.95. CODE: SCWT

THE TIME TRAVEL HANDBOOK
A Manual of Practical Teleportation & Time Travel
edited by David Hatcher Childress
In the tradition of *The Anti-Gravity Handbook* and *The Free-Energy Device Handbook*, science and UFO author David Hatcher Childress takes us into the weird world of time travel and teleportation. Not just a whacked-out look at science fiction, this book is an authoritative chronicling of real-life time travel experiments, teleportation devices and more. *The Time Travel Handbook* takes the reader beyond the government experiments and deep into the uncharted territory of early time travellers such as Nikola Tesla and Guglielmo Marconi and their alleged time travel experiments, as well as the Wilson Brothers of EMI and their connection to the Philadelphia Experiment—the U.S. Navy's forays into invisibility, time travel, and teleportation. Childress looks into the claims of time travelling individuals, and investigates the unusual claim that the pyramids on Mars were built in the future and sent back in time. A highly visual, large format book, with patents, photos and schematics. Be the first on your block to build your own time travel device!
316 PAGES. 7X10 PAPERBACK. ILLUSTRATED. $16.95. CODE: TTH

THE TESLA PAPERS
Nikola Tesla on Free Energy & Wireless Transmission of Power
by Nikola Tesla, edited by David Hatcher Childress
David Hatcher Childress takes us into the incredible world of Nikola Tesla and his amazing inventions. Tesla's rare article "The Problem of Increasing Human Energy with Special Reference to the Harnessing of the Sun's Energy" is included. This lengthy article was originally published in the June 1900 issue of *The Century Illustrated Monthly Magazine* and it was the outline for Tesla's master blueprint for the world. Tesla's fantastic vision of the future, including wireless power, anti-gravity, free energy and highly advanced solar power. Also included are some of the papers, patents and material collected on Tesla at the Colorado Springs Tesla Symposiums, including papers on: •The Secret History of Wireless Transmission •Tesla and the Magnifying Transmitter •Design and Construction of a Half-Wave Tesla Coil •Electrostatics: A Key to Free Energy •Progress in Zero-Point Energy Research •Electromagnetic Energy from Antennas to Atoms •Tesla's Particle Beam Technology •Fundamental Excitatory Modes of the Earth-Ionosphere Cavity
325 PAGES. 8X10 PAPERBACK. ILLUSTRATED. $16.95. CODE: TTP

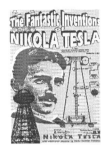

THE FANTASTIC INVENTIONS OF NIKOLA TESLA
by Nikola Tesla with additional material by David Hatcher Childress
This book is a readable compendium of patents, diagrams, photos and explanations of the many incredible inventions of the originator of the modern era of electrification. In Tesla's own words are such topics as wireless transmission of power, death rays, and radio-controlled airships. In addition, rare material on German bases in Antarctica and South America, and a secret city built at a remote jungle site in South America by one of Tesla's students, Guglielmo Marconi. Marconi's secret group claims to have built flying saucers in the 1940s and to have gone to Mars in the early 1950s! Incredible photos of these Tesla craft are included. The Ancient Atlantean system of broadcasting energy through a grid system of obelisks and pyramids is discussed, and a fascinating concept comes out of one chapter: that Egyptian engineers had to wear protective metal head-shields while in these power plants, hence the Egyptian Pharoah's head covering as well as the Face on Mars! •His plan to transmit free electricity into the atmosphere. •How electrical devices would work using only small antennas. •Why unlimited power could be utilized anywhere on earth. •How radio and radar technology can be used as death-ray weapons in Star Wars.
342 PAGES. 6X9 PAPERBACK. ILLUSTRATED. $16.95. CODE: FINT

24 hour credit card orders—call: 815-253-6390 fax: 815-253-6300
email: auphq@frontiernet.net   www.adventuresunlimitedpress.com   www.wexclub.com

MYSTIC TRAVELLER SERIES

THE MYSTERY OF EASTER ISLAND
by Katherine Routledge
The reprint of Katherine Routledge's classic archaeology book which was first published in London in 1919. The book details her journey by yacht from England to South America, around Patagonia to Chile and on to Easter Island. Routledge explored the amazing island and produced one of the first-ever accounts of the life, history and legends of this strange and remote place. Routledge discusses the statues, pyramid-platforms, Rongo Rongo script, the Bird Cult, the war between the Short Ears and the Long Ears, the secret caves, ancient roads on the island, and more. This rare book serves as a sourcebook on the early discoveries and theories on Easter Island.
432 PAGES. 6X9 PAPERBACK. ILLUSTRATED. $16.95. CODE: MEI

MYSTERY CITIES OF THE MAYA
Exploration and Adventure in Lubaantun & Belize
by Thomas Gann

First published in 1925, *Mystery Cities of the Maya* is a classic in Central American archaeology-adventure. Gann was close friends with Mike Mitchell-Hedges, the British adventurer who discovered the famous crystal skull with his adopted daughter Sammy and Lady Richmond Brown, their benefactress. Gann battles pirates along Belize's coast and goes upriver with Mitchell-Hedges to the site of Lubaantun where they excavate a strange lost city where the crystal skull was discovered. Lubaantun is a unique city in the Mayan world as it is built out of precisely carved blocks of stone without the usual plaster-cement facing. Lubaantun contained several large pyramids partially destroyed by earthquakes and a large amount of artifacts. Gann shared Mitchell-Hedges belief in Atlantis and lost civilizations (pre-Mayan) in Central America and the Caribbean. Lots of good photos, maps and diagrams.
252 PAGES. 6X9 PAPERBACK. ILLUSTRATED. $16.95. CODE: MCOM

IN SECRET TIBET
by Theodore Illion
Reprint of a rare 30s adventure travel book. Illion was a German wayfarer who not only spoke fluent Tibetan, but travelled in disguise as a native through forbidden Tibet when it was off-limits to all outsiders. His incredible adventures make this one of the most exciting travel books ever published. Includes illustrations of Tibetan monks levitating stones by acoustics.
210 PAGES. 6X9 PAPERBACK. ILLUSTRATED. $15.95. CODE: IST

DARKNESS OVER TIBET
by Theodore Illion
In this second reprint of Illion's rare books, the German traveller continues his journey through Tibet and is given directions to a strange underground city. As the original publisher's remarks said, "this is a rare account of an underground city in Tibet by the only Westerner ever to enter it and escape alive! "
210 PAGES. 6X9 PAPERBACK. ILLUSTRATED. $15.95. CODE: DOT

IN SECRET MONGOLIA
by Henning Haslund
First published by Kegan Paul of London in 1934, Haslund takes us into the barely known world of Mongolia of 1921, a land of god-kings, bandits, vast mountain wilderness and a Russian army running amok. Starting in Peking, Haslund journeys to Mongolia as part of the Krebs Expedition—a mission to establish a Danish butter farm in a remote corner of northern Mongolia. Along the way, he smuggles guns and nitroglycerin, is thrown into a prison by the new Communist regime, battles the Robber Princess and more. With Haslund we meet the "Mad Baron" Ungern-Sternberg and his renegade Russian army, the many characters of Urga's fledgling foreign community, and the last god-king of Mongolia, Seng Chen Gegen, the fifth reincarnation of the Tiger god and the "ruler of all Torguts." Aside from the esoteric and mystical material, there is plenty of just plain adventure: Haslund encounters a Mongolian werewolf; is ambushed along the trail; escapes from prison and fights terrifying blizzards; more.
374 PAGES. 6X9 PAPERBACK. ILLUSTRATED. BIB. & INDEX. $16.95. CODE: ISM

MEN & GODS IN MONGOLIA
by Henning Haslund
First published in 1935 by Kegan Paul of London, Haslund takes us to the lost city of Karakota in the Gobi desert. We meet the Bodgo Gegen, a god-king in Mongolia similar to the Dalai Lama of Tibet. We meet Dambin Jansang, the dreaded warlord of the "Black Gobi." There is even material in this incredible book on the Hi-mori, an "airhorse" that flies through the sky (similar to a Vimana) and carries with it the sacred stone of Chintamani. Aside from the esoteric and mystical material, there is plenty of just plain adventure: Haslund and companions journey across the Gobi desert by camel caravan; are kidnapped and held for ransom; witness initiation into Shamanic societies; meet reincarnated warlords; and experience the violent birth of "modern" Mongolia.
358 PAGES. 6X9 PAPERBACK. ILLUSTRATED. INDEX. $15.95. CODE: MGM

24 hour credit card orders—call: 815-253-6390 fax: 815-253-6300
email: auphq@frontiernet.net   www.adventuresunlimitedpress.com   www.wexclub.com

LOST CITIES

LOST CITIES OF ATLANTIS, ANCIENT EUROPE & THE MEDITERRANEAN
by David Hatcher Childress
Atlantis! The legendary lost continent comes under the close scrutiny of maverick archaeologist David Hatcher Childress in this sixth book in the internationally popular *Lost Cities* series. Childress takes the reader in search of sunken cities in the Mediterranean; across the Atlas Mountains in search of Atlantean ruins; to remote islands in search of megalithic ruins; to meet living legends and secret societies. From Ireland to Turkey, Morocco to Eastern Europe, and around the remote islands of the Mediterranean and Atlantic, Childress takes the reader on an astonishing quest for mankind's past. Ancient technology, cataclysms, megalithic construction, lost civilizations and devastating wars of the past are all explored in this book. Childress challenges the skeptics and proves that great civilizations not only existed in the past, but the modern world and its problems are reflections of the ancient world of Atlantis.
524 PAGES. 6X9 PAPERBACK. ILLUSTRATED WITH 100S OF MAPS, PHOTOS AND DIAGRAMS. BIBLIOGRAPHY & INDEX. $16.95. CODE: MED

LOST CITIES OF CHINA, CENTRAL INDIA & ASIA
by David Hatcher Childress
Like a real life "Indiana Jones," maverick archaeologist David Childress takes the reader on an incredible adventure across some of the world's oldest and most remote countries in search of lost cities and ancient mysteries. Discover ancient cities in the Gobi Desert; hear fantastic tales of lost continents, vanished civilizations and secret societies bent on ruling the world; visit forgotten monasteries in forbidding snow-capped mountains with strange tunnels to mysterious subterranean cities! A unique combination of far-out exploration and practical travel advice, it will astound and delight the experienced traveler or the armchair voyager.
429 PAGES. 6X9 PAPERBACK. ILLUSTRATED. FOOTNOTES & BIBLIOGRAPHY. $14.95. CODE: CHI

LOST CITIES OF ANCIENT LEMURIA & THE PACIFIC
by David Hatcher Childress
Was there once a continent in the Pacific? Called Lemuria or Pacifica by geologists, Mu or Pan by the mystics, there is now ample mythological, geological and archaeological evidence to "prove" that an advanced and ancient civilization once lived in the central Pacific. Maverick archaeologist and explorer David Hatcher Childress combs the Indian Ocean, Australia and the Pacific in search of the surprising truth about mankind's past. Contains photos of the underwater city on Pohnpei; explanations on how the statues were levitated around Easter Island in a clockwise vortex movement; tales of disappearing islands; Egyptians in Australia; and more.
379 PAGES. 6X9 PAPERBACK. ILLUSTRATED. FOOTNOTES & BIBLIOGRAPHY. $14.95. CODE: LEM

ANCIENT TONGA
& the Lost City of Mu'a
by David Hatcher Childress
Lost Cities series author Childress takes us to the south sea islands of Tonga, Rarotonga, Samoa and Fiji to investigate the megalithic ruins on these beautiful islands. The great empire of the Polynesians, centered on Tonga and the ancient city of Mu'a, is revealed with old photos, drawings and maps. Chapters in this book are on the Lost City of Mu'a and its many megalithic pyramids, the Ha'amonga Trilithon and ancient Polynesian astronomy, Samoa and the search for the lost land of Havai'iki, Fiji and its wars with Tonga, Rarotonga's megalithic road, and Polynesian cosmology. Material on Egyptians in the Pacific, earth changes, the fortified moat around Mu'a, lost roads, more.
218 PAGES. 6X9 PAPERBACK. ILLUSTRATED. COLOR PHOTOS. BIBLIOGRAPHY. $15.95. CODE: TONG

ANCIENT MICRONESIA
& the Lost City of Nan Madol
by David Hatcher Childress
Micronesia, a vast archipelago of islands west of Hawaii and south of Japan, contains some of the most amazing megalithic ruins in the world. Part of our *Lost Cities* series, this volume explores the incredible conformations on various Micronesian islands, especially the fantastic and little-known ruins of Nan Madol on Pohnpei Island. The huge canal city of Nan Madol contains over 250 million tons of basalt columns over an 11 square-mile area of artificial islands. Much of the huge city is submerged, and underwater structures can be found to an estimated 80 feet. Islanders' legends claim that the basalt rocks, weighing up to 50 tons, were magically levitated into place by the powerful forefathers. Other ruins in Micronesia that are profiled include the Latte Stones of the Marianas, the menhirs of Palau, the megalithic canal city on Kosrae Island, megaliths on Guam, and more.
256 PAGES. 6X9 PAPERBACK. ILLUSTRATED. COLOR PHOTOS. BIBLIOGRAPHY. $16.95. CODE: AMIC

24 hour credit card orders—call: 815-253-6390 fax: 815-253-6300
email: auphq@frontiernet.net   www.adventuresunlimitedpress.com   www.wexclub.com

LOST CITIES

TECHNOLOGY OF THE GODS
The Incredible Sciences of the Ancients
by David Hatcher Childress
Popular *Lost Cities* author David Hatcher Childress takes us into the amazing world of ancient technology, from computers in antiquity to the "flying machines of the gods." Childress looks at the technology that was allegedly used in Atlantis and the theory that the Great Pyramid of Egypt was originally a gigantic power station. He examines tales of ancient flight and the technology that it involved; how the ancients used electricity; megalithic building techniques; the use of crystal lenses and the fire from the gods; evidence of various high tech weapons in the past, including atomic weapons; ancient metallurgy and heavy machinery; the role of modern inventors such as Nikola Tesla in bringing ancient technology back into modern use; impossible artifacts; and more.
356 PAGES. 6x9 PAPERBACK. ILLUSTRATED. BIBLIOGRAPHY. $16.95. CODE: TGOD

VIMANA AIRCRAFT OF ANCIENT INDIA & ATLANTIS
by David Hatcher Childress, introduction by Ivan T. Sanderson

Did the ancients have the technology of flight? In this incredible volume on ancient India, authentic Indian texts such as the *Ramayana* and the *Mahabharata* are used to prove that ancient aircraft were in use more than four thousand years ago. Included in this book is the entire Fourth Century BC manuscript *Vimaanika Shastra* by the ancient author Maharishi Bharadwaaja, translated into English by the Mysore Sanskrit professor G.R. Josyer. Also included are chapters on Atlantean technology, the incredible Rama Empire of India and the devastating wars that destroyed it. Also an entire chapter on mercury vortex propulsion and mercury gyros, the power source described in the ancient Indian texts. Not to be missed by those interested in ancient civilizations or the UFO enigma.
334 PAGES. 6x9 PAPERBACK. ILLUSTRATED. $15.95. CODE: VAA

LOST CONTINENTS & THE HOLLOW EARTH
I Remember Lemuria and the Shaver Mystery
by David Hatcher Childress & Richard Shaver

Lost Continents & the Hollow Earth is Childress' thorough examination of the early hollow earth stories of Richard Shaver and the fascination that fringe fantasy subjects such as lost continents and the hollow earth have had for the American public. Shaver's rare 1948 book *I Remember Lemuria* is reprinted in its entirety, and the book is packed with illustrations from Ray Palmer's *Amazing Stories* magazine of the 1940s. Palmer and Shaver told of tunnels running through the earth—tunnels inhabited by the Deros and Teros, humanoids from an ancient spacefaring race that had inhabited the earth, eventually going underground, hundreds of thousands of years ago. Childress discusses the famous hollow earth books and delves deep into whatever reality may be behind the stories of tunnels in the earth. Operation High Jump to Antarctica in 1947 and Admiral Byrd's bizarre statements, tunnel systems in South America and Tibet, the underground world of Agartha, the belief of UFOs coming from the South Pole, more.
344 PAGES. 6x9 PAPERBACK. ILLUSTRATED. $16.95. CODE: LCHE

LOST CITIES OF NORTH & CENTRAL AMERICA
by David Hatcher Childress

Down the back roads from coast to coast, maverick archaeologist and adventurer David Hatcher Childress goes deep into unknown America. With this incredible book, you will search for lost Mayan cities and books of gold, discover an ancient canal system in Arizona, climb gigantic pyramids in the Midwest, explore megalithic monuments in New England, and join the astonishing quest for lost cities throughout North America. From the war-torn jungles of Guatemala, Nicaragua and Honduras to the deserts, mountains and fields of Mexico, Canada, and the U.S.A., Childress takes the reader in search of sunken ruins, Viking forts, strange tunnel systems, living dinosaurs, early Chinese explorers, and fantastic lost treasure. Packed with both early and current maps, photos and illustrations.
590 PAGES. 6x9 PAPERBACK. ILLUSTRATED. FOOTNOTES & BIBLIOGRAPHY. $14.95. CODE: NCA

LOST CITIES & ANCIENT MYSTERIES OF SOUTH AMERICA
by David Hatcher Childress
Rogue adventurer and maverick archaeologist David Hatcher Childress takes the reader on unforgettable journeys deep into deadly jungles, high up on windswept mountains and across scorching deserts in search of lost civilizations and ancient mysteries. Travel with David and explore stone cities high in mountain forests and hear fantastic tales of Inca treasure, living dinosaurs, and a mysterious tunnel system. Whether he is hopping freight trains, searching for secret cities, or just dealing with the daily problems of food, money, and romance, the author keeps the reader spellbound. Includes both early and current maps, photos, and illustrations, and plenty of advice for the explorer planning his or her own journey of discovery.
381 PAGES. 6x9 PAPERBACK. ILLUSTRATED. FOOTNOTES. BIBLIOGRAPHY. $14.95. CODE: SAM

LOST CITIES & ANCIENT MYSTERIES OF AFRICA & ARABIA
by David Hatcher Childress

Across ancient deserts, dusty plains and steaming jungles, maverick archaeologist David Childress continues his world-wide quest for lost cities and ancient mysteries. Join him as he discovers forbidden cities in the Empty Quarter of Arabia; "Atlantean" ruins in Egypt and the Kalahari desert; a mysterious, ancient empire in the Sahara; and more. This is the tale of an extraordinary life on the road: across war-torn countries, Childress searches for King Solomon's Mines, living dinosaurs, the Ark of the Covenant and the solutions to some of the fantastic mysteries of the past.
423 PAGES. 6x9 PAPERBACK. ILLUSTRATED. FOOTNOTES & BIBLIOGRAPHY. $14.95. CODE: AFA

24 hour credit card orders—call: 815-253-6390 fax: 815-253-6300
email: auphq@frontiernet.net   www.adventuresunlimitedpress.com   www.wexclub.com

One Adventure Place
P.O. Box 74
Kempton, Illinois 60946
United States of America
Tel.: 815-253-6390 • Fax: 815-253-6300
Email: auphq@frontiernet.net
http://www.adventuresunlimitedpress.com
or www.wexclub.com/aup

ORDERING INSTRUCTIONS

✓ Remit by USD$ Check, Money Order or Credit Card
✓ Visa, Master Card, Discover & AmEx Accepted
✓ Prices May Change Without Notice
✓ 10% Discount for 3 or more Items

SHIPPING CHARGES

United States

✓ Postal Book Rate { $3.00 First Item / 50¢ Each Additional Item
✓ Priority Mail { $4.00 First Item / $2.00 Each Additional Item
✓ UPS { $5.00 First Item / $1.50 Each Additional Item
NOTE: UPS Delivery Available to Mainland USA Only

Canada

✓ Postal Book Rate { $4.00 First Item / $1.00 Each Additional Item
✓ Postal Air Mail { $6.00 First Item / $2.00 Each Additional Item
✓ Personal Checks or Bank Drafts MUST BE USD$ and Drawn on a US Bank
✓ Canadian Postal Money Orders OK
✓ Payment MUST BE USD$

All Other Countries

✓ Surface Delivery { $7.00 First Item / $2.00 Each Additional Item
✓ Postal Air Mail { $13.00 First Item / $8.00 Each Additional Item
✓ Payment MUST BE USD$
✓ Checks and Money Orders MUST BE USD$ and Drawn on a US Bank or branch.
✓ Add $5.00 for Air Mail Subscription to Future *Adventures Unlimited* Catalogs

SPECIAL NOTES

✓ RETAILERS: Standard Discounts Available
✓ BACKORDERS: We Backorder all Out-of-Stock Items Unless Otherwise Requested
✓ PRO FORMA INVOICES: Available on Request
✓ VIDEOS: NTSC Mode Only. Replacement only.
✓ For PAL mode videos contact our other offices:

European Office:
Adventures Unlimited, Panewaal 22,
Enkhuizen, 1600 AA, The Netherlands
http: www.adventuresunlimited.nl
Check Us Out Online at:
www.adventuresunlimitedpress.com

Please check: ☑

☐ This is my first order ☐ I have ordered before ☐ This is a new address

Name	
Address	
City	
State/Province	Postal Code
Country	
Phone day	Evening
Fax	

Item Code	Item Description	Price	Qty	Total

Please check: ☑

☐ Postal-Surface
☐ Postal-Air Mail (Priority in USA)
☐ UPS (Mainland USA only)

Subtotal ➠	
Less Discount-10% for 3 or more items ➠	
Balance ➠	
Illinois Residents 6.25% Sales Tax ➠	
Previous Credit ➠	
Shipping ➠	
Total (check/MO in USD$ only)➠	

☐ Visa/MasterCard/Discover/Amex

Card Number

Expiration Date

10% Discount When You Order 3 or More Items!

Comments & Suggestions

Share Our Catalog with a Friend